Polymer Thin Films: From Fundamentals to Applications (Second Edition)

Polymer Thin Films: From Fundamentals to Applications (Second Edition)

Guest Editor

Mohor Mihelčič

MDPI

Basel • Beijing • Wuhan • Barcelona • Belgrade • Novi Sad • Cluj • Manchester

Guest Editor
Mohor Mihelčič
Laboratory for Experimental
Mechanics
University of Ljubljana
Ljubljana
Slovenija

Editorial Office
MDPI AG
Grosspeteranlage 5
4052 Basel, Switzerland

This is a reprint of the Special Issue, published open access by the journal *Coatings* (ISSN 2079-6412), freely accessible at: https://www.mdpi.com/journal/coatings/special_issues/3MD2L1S1MG.

For citation purposes, cite each article independently as indicated on the article page online and as indicated below:

Lastname, A.A.; Lastname, B.B. Article Title. *Journal Name* **Year**, *Volume Number*, Page Range.

ISBN 978-3-7258-6011-1 (Hbk)
ISBN 978-3-7258-6012-8 (PDF)
https://doi.org/10.3390/books978-3-7258-6012-8

Cover image courtesy of Mohor Mihelčič

Contents

About the Editor

Mohor Mihelčič

Mohor Mihelčič is a research associate at the Faculty of Mechanical Engineering, University of Ljubljana, Slovenia. He received his Ph.D. in Nanoscience and Nanotechnology at Jožef Stefan International Postgraduate School, Slovenia in 2015. Mohor has co-authored 44 scientific publications and 58 conferences, as well as 2 patents. His research focuses on the preparation and characterization of multifunctional coatings and polymers such as spectrally selective thickness sensitive (TSSS) and thickness insensitive (TISS) coatings, antimicrobial, hydrophobic, photocatalytic, thermo- and electrochromic coatings.

coatings MDPI

rticle

The Addition of MoO_3 or SiO_2 Nano-/Microfillers Thermally Stabilized and Mechanically Reinforce the PVDF-HFP/PVP Polymer Composite Thin Films

rška Gradišar Centa [1], Anja Pogačnik Krajnc [2,3], Lidija Slemenik Perše [1], Matic Šobak [1] nd Mohor Mihelčič [1,*]

[1] Faculty of Mechanical Engineering, University of Ljubljana, Aškerčeva 6, 1000 Ljubljana, Slovenia; urska.gradisarcenta@fs.uni-lj.si (U.G.C.); lidija.slemenik.perse@fs.uni-lj.si (L.S.P.); matic.sobak@fs.uni-lj.si (M.Š.)
[2] Institut Jozef Stefan, Jamova 39, 1000 Ljubljana, Slovenia; anja.pogacnik@ijs.si
[3] Faculty of Mathematics and Physics, University of Ljubljana, 1000 Ljubljana, Slovenia
* Correspondence: mohor.mihelcic@fs.uni-lj.si

Abstract: The properties of thin polymer films are influenced by the size of the fillers, their morphology, the surface properties and their distribution/interaction in the polymer matrix. In this work, thin polymer composite films with MoO_3 or SiO_2 nano and micro fillers in PVDF-HFP/PVP polymer matrix were successfully fabricated using the solvent casting method. The effects of different types, sizes and morphologies of the inorganic fillers on the crystallization of the PVDF-HFP polymer were investigated, as well as the effects on the thermal and mechanical properties of the composites. Scanning electron microscopy, ATR-FTIR spectroscopy, differential scanning calorimetry, nanoindentation and uniaxial mechanical tests were used for characterization. The results showed that MoO_3 nanowires thermally stabilized the polymer matrix, induced crystallization of the PVDF-HFP polymer in all three polymorphs (α-, β-, γ-phase) and formed a geometrical network in the polymer matrix, resulting in the highest elastic moduli, hardness and Young's modulus.

Keywords: polymer composites; crystallization; PVP; PVDF-HFP; mechanical properties; thermal properties

1. Introduction

Polymer composites are promising new engineering materials with improved thermal, mechanical, and surface properties. Various fillers have been used to improve different properties of polymers, i.e., carbon nanotubes [1], glass fibers [2], cellulose [3], silica-SiO_2 particles [4], SiC and Al_2O_3 nanowires [5], and others. The particle size and morphology of the fillers, their orientation, dispersion, and concentration in polymer matrix, the adhesion between the polymer matrix and the filler and the interphase properties have already been proved to have a major influence on the final properties of the composites. It has already been shown that the particle size has a great influence on the mechanical properties, such as Young's modulus [6], which is due to the different degree of interactions between the filler and the polymer matrix. For spherical Al_2O_3 particles, it was found that the addition of microparticles has only a small effect on the Young's modulus, while the Young's modulus increases significantly with the addition of nanoparticles [7].

The polymer composites based on polyvinylidene fluoride (PVDF) or poly (vinylidene fluoride-co-hexafluoropropylene) (PVDF-HFP) polymers are used in various attractive applications, i.e., as flexible, biocompatible nanofibers with superior piezoelectric/ferroelectric properties in the field of bone tissue regeneration [8], flexible electronic skins [9], nanogenerators [10], or intelligent and smart packaging material with water vapor impermeability [11]. Additionally, magnetoactive polymer composites with $CoFe_2O_4$ nanoparticles in a polyvinyl alcohol (PVA) polymer matrix are being investigated for applications in the fields of remote sensing, soft robotics, electronics, and biomedicine [12]. Another useful application is flexible energy storage devices such as supercapacitors for extreme conditions [13]. For the prediction of electric field distribution in the PVDF based nanocomposites with various fillers, the researchers in the last years have used COMSOL software [14].

The increase in crystallinity affects not only the morphological characteristics but also the mechanical and thermal properties [15]. Factors such as particle size and particle-particle/particle-matrix interactions influence the crystallization of polymers with crystalline characters. The PVDF-HFP polymer is a semi-crystalline polymer that can crystallize into α-, β-, and γ- phases, depending on the temperature, processing conditions and fillers/additives used. The doping of PVDF-HFP polymer with fillers, most commonly organic particles [16] or clay minerals [17], induce the crystallization in the most polar β-phase, which was also formed in the case of mechanical deformation of non-polar α-phase [18]. Especially, the addition of water-soluble polyvinylpyrrolidone (PVP) polymer induces the crystallization of PVDF-HFP polymer in β-phase, and even more, PVP increases its stabilization [19].

During the evaporation process of the solvent (solidification) the molecular chains adjacent to the particles contract earlier than the free polymer chains, forming a high-density zone with higher elastic modulus around the particles [20]. The polymer matrix as the main component of the polymer composite carries the mechanical load, while the dispersed particles primarily increase the material's elastic modulus [21]. Moreover, the incorporation of nanoparticles into the polymer matrix significantly improves the structure of the cross-linked network between the polymer matrix and the filler, as they serve as cross-linking attractors, which is particularly observed in the case of large deformations [22].

For the formation of high-strength composites, the adhesion between the filler and the polymer matrix is a very important parameter that depends on the geometry of the filler, the chemical structure of the polymer, the regularity of the molecular structure, the conformational properties, and the branching of the polymer chains [21]. Furthermore, Woigk et al. reported that interfacial properties correlate with strength in both longitudinal and transverse directions in a nanocomposite based on a polylactic acid (PLLA) polymer reinforced with natural fibers [23].

MoO_3 nanowires are used in a diverse range of applications, i.e., they can enhance the antimicrobial activity of polymer nanocomposite films [19,24,25]. However, the fabrication of these nanoparticles is limited due to its complex nature and high cost. On the other hand, SiO_2 filler, which has not yet been used in the PVDF-HFP/PVP polymer matrix, is widely used in the pharmaceutical industry. This nanomaterial could be a promising candidate to induce antifouling and antimicrobial properties in the polymer matrix. In our previous study, we showed that the SiO_2 nanoparticles in PVDF-HFP/PVP based polymer nanocomposite reduced the adhesion of *S. aureus* and *E. coli* bacterial cells after 24 h of incubation [26]. Since SiO_2 filler is more affordable, its use would improve the availability and application possibilities of such polymer composites.

In this research, we investigated the effects of adding morphologically different inorganic fillers, such as MoO_3 and SiO_2 on the thermal, structural and mechanical properties of polymer composites based on PVDF-HFP and PVP polymer. Furthermore, we compared how different filler sizes affect crystallization, the presence of the individual crystal phases and the interactions between fillers and the polymer matrix, which also determines the mechanical properties of polymer composites.

2. Materials and Methods

The polymer composites were prepared using poly (vinylidene fluoride-co-hexafluoropropylene) (PVDF-HFP) (Sigma Aldrich, St. Louis, MO, USA) and polyvinylpyrrolidone (PVP) polymers (Sigma Aldrich, St. Louis, MO, USA) polymer. MoO_3 nanowires (Nanotul Ltd, Ljubljana, Slovenia), MoO_3 microparticles (Sigma Aldrich, St. Louis, MO, USA), SiO_2 nanoparticles (W.R. Grace, Columbia, MD, USA), and SiO_2 microparticles (W.R. Grace,

Columbia, MD, USA) were used as fillers. For each polymer composite, both polymers were dissolved separately in dimethylformamide (DMF, Carlo Erba Reagents, Cornaredo, Italy) and mixed for 4 h at 80 °C and 300 rpm on a magnetic stirrer. The fillers were then added to the PVP polymer solution, and the dispersion was mixed for further 2 h under the same conditions. In the last step of the mixing process, the PVDF-HFP polymer solution was added to the dispersion and mixed for 2 additional hours. After the mixing process, the polymer composite solutions were casted on a Teflon plate and dried at 80 °C for 2 h. The composition of the polymer composites was constant, with the mass ratio of PVDF-HFP, PVP and filler being 69 wt. %, 23 wt. %, and 8 wt. %, respectively. The prepared polymer composites are shown in Figure 1.

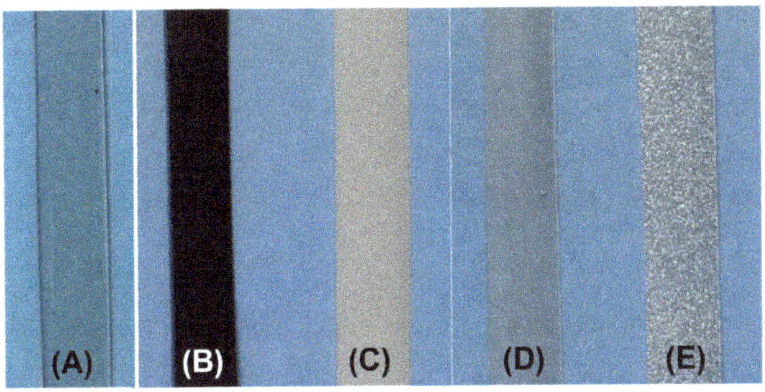

Figure 1. Samples of (**A**) polymer matrix and polymer composites: with (**B**) MoO_3 nanowires, (**C**) MoO_3 microparticles, (**D**) SiO_2 nanoparticles, (**E**) SiO_2 microparticles for characterization.

The surface topography of polymer composite thin films was investigated with a scanning electron microscope Verios 4G HP, Thermo Fisher (Waltham, MA, USA). The images were generated from the signal of secondary electrons at a working distance of about 4 mm and an accelerated voltage of 3–5 kV. The nano- and micro-particles in powder form were placed on an adhesive carbon tape fixed in the sample holder and blown with argon to remove the excess powder. The polymer composite thin films were placed on an adhesive carbon tape and sputtered with a 10 nm carbon layer.

The ATR-FTIR spectra were recorded at room temperature with a FTIR spectrometer (Perkin Elmer, Spectrum 65, Waltham, MA, USA), equipped with a single reflection diamond crystal. The ATR-FTIR spectra of free-standing polymer composite thin films were recorded in a range from 4000 to 600 cm^{-1}, with 64 scans per spectrum and a resolution of 4 cm^{-1}.

Differential scanning calorimetry-DSC was performed using the Q2500 calorimeter (TA Instruments, New Castle, DE, USA). Heat–cool–reheat tests were performed in a nitrogen atmosphere in the temperature range from −90 °C to 180 °C with heating/cooling rates of 10 °C/min. The thermal properties of the polymer composites, i.e., the melting and crystallization temperature, the glass transition temperature and the enthalpies were evaluated using the TRIOS software. From these data, the degree of crystallinity of the polymer composites was calculated using the following equation:

$$\chi_c = \frac{\Delta H_f}{\Delta H_f^0 \cdot f} \cdot 100\%$$

where X_c is the degree of crystallinity, ΔH_f is the fusion enthalpy, ΔH_f^0 is the fusion enthalpy for 100% crystallinity (for PVDF-HFP polymer, the value of 104.7 J/g was used [27]) and f is the weight fraction of PVDF-HFP polymer in composite.

The elastic modulus and hardness on the surface of polymer composite films were determined using nanoindentation. The free-standing polymer composite films were placed on the glass coated with two-component glue and fixed on a holder. The continuous stiffness measurements (CSM) were performed by Nanoindenter G200 XP instrument (Agilent Technologies, Santa Clara, CA, USA) using a standard three-sided pyramidal Berkovich probe at the tip oscillation frequency of 45 Hz and 2 nm harmonic amplitude. For each sample, the 36 indents were performed. The values of elastic modulus and hardness were determined at a depth of 1200 and 2400 nm and Poisson's ratio of 0.4 for PVDF-HFP [28].

Uniaxial mechanical tests were carried out by the modular compact rheometer MCR 702 (Anton Paar, Graz, Austria) with samples, which were 5 mm wide, 40 mm long and 150 μm thick. The measurements were carried out at room temperature with pre-applied force of 0.3 N in three repetitions for each polymer composite film. Average stress-strain curves were presented as the result of the characterization. Young's modulus was calculated as a slope of stress-strain curves in a linear range.

3. Results and Discussion

3.1. Morphology

The properties of polymer composites depend on the morphology, size and distribution of the filler within the polymer matrix. Therefore, the MoO_3 and SiO_2 fillers, which differ in size and morphology, were used to produce the polymer composites. The morphology of all four polymer composite fillers is shown in Figure 2. The MoO_3 nanowires (with a diameter of about 200 nm and a length of up to 3 μm), which grow in an ortho-rhombic crystal structure, have a large specific surface area (12.06 ± 0.05 m^2/g) [29] (Figure 2a). The commercially available MoO_3 microparticles also exhibit an orthorhombic crystal structure, with an average single crystal size of about 10 μm [29]. Since they show a tendency to form larger agglomerates (Figure 2b), they have smaller specific surface area of 0.27 ± 0.05 m^2/g [29]. The mesoporous SiO_2 nanoparticles, which tend to form agglomerates (Figure 2c), have the largest specific surface area of (345.55 ± 1.26 m^2/g) [30]. In contrast, the SiO_2 microparticles were the largest filler particles with irregular shapes and an average particle size of about 150 μm (Figure 2d). They have a specific surface area of 305.51 ± 1.23 m^2/g [31].

Figure 2. SEM images of nano- and micro-particles used: (**a**) MoO_3 nanowires, (**b**) MoO_3 microparticles, (**c**) SiO_2 nanoparticles, (**d**) SiO_2 microparticles.

The surface topography of PVDF-HFP/PVP polymer matrix and polymer composite films with MoO_3 and SiO_2 nano- and micro-particles is shown in Figure 3. Partially circular polymer domains can be observed on the surface of the polymer matrix (Figure 3a) and the polymer composite filled with MoO_3 particles (Figure 3b,c). The viscoelastic properties of PVDF-HFP and PVP polymer solution were different, which leads to the formation of phase separation structures in polymer composites during the film formation process. The PVP domains with fillers in the PVDF-HFP polymer matrix are formatted, which appear as spherical structures on the surface. The incorporation of PVP reduces the overall crystallinity of PVDF-HFP, which can promote the formation of a more amorphous and nanostructured surface. The interaction between PVP and PVDF-HFP induces the self-assembly of PVP into spherical nanostructures with an average diameter of 200 to 500 nm [19]. However, when MoO_3 nano- or microparticles were added, these spherical structures increase in size to ~2 µm. The MoO_3 nanowires were homogeneously dispersed on the surface of the polymer composite (Figure 3b). In addition, the polymer domains were also visible on the surface of the polymer composite with MoO_3 microparticles (Figure 3c), where the rough surface was observed because of the larger dimension of the fillers, which were, however, relatively homogeneously dispersed. The coatings with the SiO_2 nanoparticles consist of agglomerated particles (Figure 3d), which are visible on the surface due to the very low density of the nanoparticles which prevent their sedimentation in the polymer matrix. In contrast, when SiO_2 microparticles were added, the particles did not agglomerate, but larger microparticles are visible on the surface (Figure 3e).

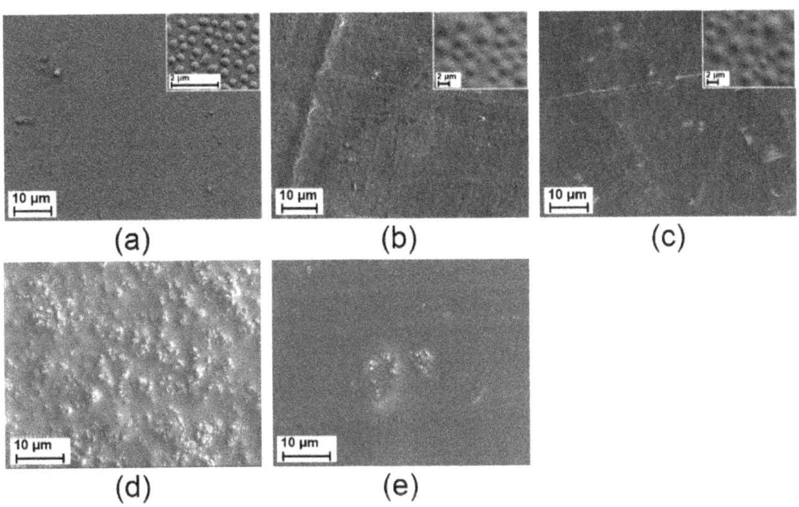

Figure 3. SEM images of the surface of (**a**) PVDF-HFP/PVP polymer matrix and polymer composite thin films and with the addition of: (**b**) MoO_3 nanowires, (**c**) MoO_3 microparticles, (**d**) SiO_2 nanoparticles, (**e**) SiO_2 microparticles. The enlargement of the polymer domains is shown in the insert.

3.2. Structural Properties

The structure of polymer composites, in particular the interactions between MoO_3 or SiO_2 nano/micro fillers and the polymer matrix, was investigated by ATR-FTIR spectroscopy. The vibrational modes of the chemical bonds provide information about the crystalline phases in polymer composites and the dispersity of the fillers in the polymer matrix, which is reflected in a shift of peak intensity and position.

The ATR-FTIR spectra of PVDF-HFP/PVP polymer matrix and polymer composites with MoO_3 and SiO_2 nano/micro fillers are shown in Figure 4. FTIR spectroscopy shows the presence of mixed crystal phases in the polymer composites, however the ratio between the phases was different depending on the type of filler.

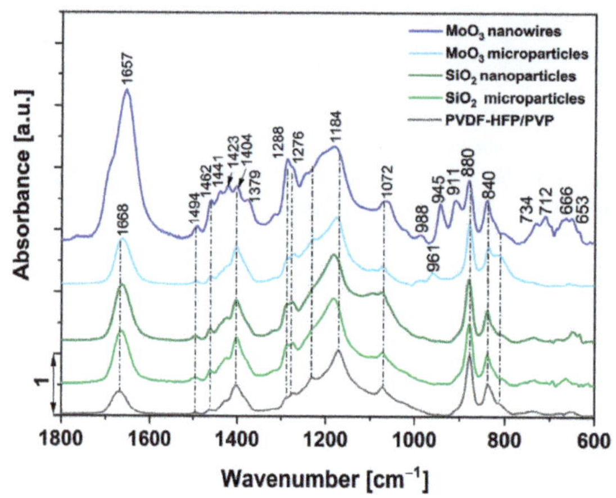

Figure 4. ATR-FTIR curves for PVDF-HFP/PVP polymer matrix and polymer composites with MoO_3 nanowires, MoO_3 microparticles, SiO_2 nanoparticles, SiO_2 microparticles.

In the PVDF-HFP/PVP polymer matrix film, the peaks corresponding to the electroactive phases ($\beta + \gamma$) of PVDF-HFP were observed at 840 cm^{-1}. Additionally, the combination of the $\alpha + \beta + \gamma$ phases were observed as a collective peak at 1176 cm^{-1} (symmetric stretching of –C–C–C-bond [32]) and 1401 cm^{-1} in FTIR spectra and at 880 cm^{-1} and 1403 cm^{-1} in Raman spectra [33]. These high intensity peaks (at 840 cm^{-1}, 880 cm^{-1} and 1403 cm^{-1}) correspond to rocking of CH_2 and asymmetric stretching of CF_2 groups, symmetric stretching of CF_2 and C-C groups [34], and CH_2 swinging [35]. The clearly visible peak at 1234 cm^{-1} also in the PVDF-HFP/PVP polymer matrix film is the exclusive peak, typically used for the identification of γ phase [36]. A strong peak at about 1668 cm^{-1} was attributed to C=O and C-N stretching in the PVP polymer [37].

The bands observed at 1423 cm^{-1}, 1462 cm^{-1} and 1494 cm^{-1} correspond to the bending of the CH_2 group in the PVP polymer [34,37] which was present in all spectra. The peak at 1072 cm^{-1} is a composition of two overlapping peaks of PVDF and SiO_2. In the case of the addition of SiO_2 filler, the peak at 1060–1070 cm^{-1} [38] is characteristic of the Si-O bond and has caused the broader peak. However, this peak was present also in all spectra of nanocomposite, which is due to the out-of-plane vibration of CF_3 bound in the PVDF-HFP polymer [39]. Furthermore, the shift to a higher wavenumber of the peak at 1184 cm^{-1} was observed in the samples with SiO_2 addition.

The polymorphic crystalline polymer PVDF-HFP can exhibit different mechanical properties depending on the presence of a crystal phase in the composite material. The characteristic peaks for the γ-phase of the PVDF-HFP polymer at 811 cm^{-1}, 1234 cm^{-1}, and 1429 cm^{-1} are present in spectra with MoO_3/SiO_2 microparticles and SiO_2 nanoparticles. However, in the composites with SiO_2 microparticles and MoO_3 nanoparticles, the absorption peak at 614 cm^{-1} indicates the formation of a small amount of α-phase. The increase in α-phase was observed in the composite with MoO_3 nanowires, while characteristic peaks at 1209 cm^{-1} and 1423 cm^{-1} were found in the spectra. Moreover, the highest intensity of the peaks at 1276 cm^{-1} (typically an exclusive peak for β phase) and 840 cm^{-1} also indicates the presence of a larger amount of β-phase in the composite with MoO_3 nanowires. Meanwhile, the peak corresponding to the CF_2 symmetrical stretching at 1424 cm^{-1} [40] has the highest intensity in the case of MoO_3 nanowires addition. According to these results, we can expect the best mechanical properties of polymer nanocomposite with MoO_3.

The SiO_2 nano/microparticles induced the crystallization of the PVDF-HFP polymer in the γ-phase, while SiO_2 microparticles promote the formation of the α-phase. The γ-phase is usually formed by the addition of the fillers such as clays and by solid phase transformation from the non-polar α-phase to the polar γ-phase [41]. In addition, the low evaporation rate of DMF in the oven (at 80 °C) reduces the mobility of the polymer chains, favoring the formation of the β-phase [42]. Moreover, nucleation in the γ-phase is also a consequence of the rapid evaporation of the solvent [42]. Furthermore, the mesoporous SiO_2 nanorods induced the formation of the polar β-phase of PVDF due to the intermolecular hydrogen interactions in the interphase [43], as observed in the characteristic FTIR peak at 1276 cm^{-1}. On the other hand, not only the filler but also the PVP polymer has an influence on the crystallization of PVDF; PVP also stabilized the β-phase of PVDF [19].

3.3. Thermal Properties

The DSC curves of the polymer matrix and polymer composites with MoO_3 and SiO_2 nano/micro fillers are shown in Figure 5, where Figure 5a shows the reheating curves and Figure 5b the cooling curves. The values of the glass transition temperatures (T_g), melting temperatures (T_m) and enthalpy of fusion (ΔH) were determined from the reheating curves, while the crystallization temperatures (T_c) and enthalpies were determined from the cooling curves. All temperatures and enthalpies are listed in Table 1. The lowest glass transition temperature was determined for PVDF-HFP/PVP polymer matrix, which was 6.5 °C lower compared to the pure PVDF-HFP film [18]. When fillers were added to the polymer matrix, the polymer composite with MoO_3 nanowires exhibited the lowest glass transition temperature (−38.42 °C), while the phase transitions were similar for the other composites. The melting temperature of PVDF-HFP/PVP polymer blend was ~12 °C lower compared to the pure PVDF-HFP film (T_m = 142.00 °C). This temperature shift is due to the addition of PVP polymer, as it acts as a plasticizer and can absorb moisture at room conditions [18]. The addition of MoO_3 microparticles and SiO_2 nanoparticles also lowered the melting temperature of the composites by about 2–6 °C compared to the PVDF-HFP/PVP polymer matrix. The highest melting enthalpy was observed with the addition of SiO_2 nanoparticles, followed by MoO_3 nanowires and SiO_2 microparticles, while the lowest value was observed for the polymer composite with MoO_3 microparticles. Moreover, the crystallinity of the polymer composites was reduced in the same order from 12.72% to 12.28%, 11.00% and 5.96%, respectively.

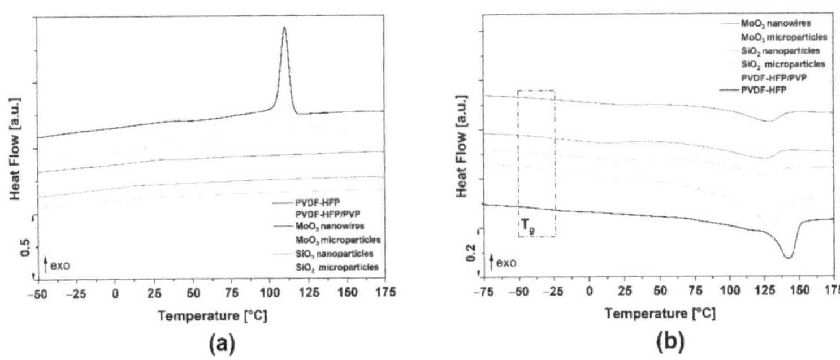

Figure 5. DSC curves for polymer composites: (**a**) second heating step, (**b**) cooling step.

Table 1. Parameters of DSC curves for pure PVDF-HFP, polymer matrix (PVDF-HFP/PVP) and polymer composites with MoO_3 and SiO_2 nano- and micro-particles.

Polymer Films with	T_g [°C]	T_m [°C]	ΔH_m [J/g]	T_c [°C]	ΔH_c [J/g]	X_C [%]
PVDF-HFP	−35.66	142.00	36.42	110.39	32.60	34.79
PVDF-HFP/PVP	−40.33	130.19	30.78	98.49	10.09	40.01
MoO_3 nanowires	−38.42	127.50	8.87	29.60	2.31	12.28
MoO_3 microparticles	−30.96	124.41	4.31	22.53	0.42	5.96
SiO_2 nanoparticles	−32.82	124.36	9.19	15.46	0.93	12.72
SiO_2 microparticles	−32.95	128.07	7.94	20.96	0.73	11.00

3.4. Mechanical Properties

The mechanical properties of the polymer composite surfaces were evaluated by nanoindentation. The elastic modulus (Figure 6a) and hardness (Figure 6b) are shown because of the measurements. Compared to the pure polymer blend PVDF-HFP/PVP, the addition of MoO_3 microparticles and SiO_2 fillers did not significantly change the elastic modulus and hardness. The elastic modulus of the polymer composite with MoO_3 microparticles was 1.85 ± 0.02 GPa, with SiO_2 nanoparticles 1.73 ± 0.01 GPa, and with SiO_2 microparticles 1.86 ± 0.02 GPa. The highest hardness (0.16 ± 0.001 GPa) of the polymer composite surface was observed in the case of the addition of MoO_3 nanowires, which was due to the highest stiffness of the polymer chains fixed in the nanowire network. However, the other fillers slightly increased the hardness of the pure polymer matrix surface (0.06 ± 0.001 GPa [18]), with 0.076 ± 0.001 GPa for MoO_3 microparticles, 0.072 ± 0.0005 GPa for SiO_2 nanoparticles and 0.088 ± 0.001 GPa for SiO_2 microparticles, respectively.

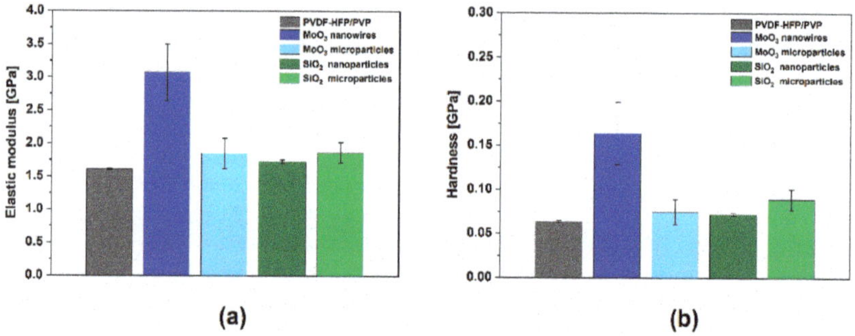

Figure 6. Nanoindentation: (**a**) elastic modulus and (**b**) hardness of PVDF-HFP/PVP polymer blend and polymer composites with MoO_3 and SiO_2 nano- and micro-particles.

In addition, Figure 7a shows the stress-strain curves of the polymer composites examined. The test results indicate that the fillers in nano-dimensions significantly enhance the mechanical properties with MoO_3 nanowires achieving the highest extensional stress values at the same strain, followed by SiO_2 nanoparticles. However, when microparticles were added, the mechanical behavior of the polymer composite was almost independent of the particle type. The same trend was observed for the Young's modulus (Figure 7b), which describes the mechanical properties of the material and indicates stiffness and resistance to deformation by external forces. The highest stiffness of the polymer composite was achieved with the addition of MoO_3 nanowires, attributed to the largest number of internal connections or bonds between the filler and the polymer chains. The Young's modulus value for the PVDF-HFP/PVP polymer matrix was 1347.6 ± 8.3 MPa, while it was lower for the polymer composite with MoO_3 nanowires (1544.1 ± 8.8 MPa), but higher when SiO_2 nanoparticles were added (1199.7 ± 7.6 MPa). This could be due to the high degree of crystallinity of the PVDF-HFP/PVP polymer matrix, which makes the polymer film stiffer. The polymer composites with the addition of MoO_3 or SiO_2 microparticles exhibited comparable Young's modulus values (882.5 ± 6.9 MPa for MoO_3 and 893.3 ± 3.5 MPa for SiO_2).

While the addition of fillers generally improves thermal and mechanical properties, the combined effects of the matrix type, nanoparticle size, and dispersion quality are also very important [42]. Due to synergy of load transfer from the matrix to reinforcement particles in nano-dimensions the nanocomposite with MoO_3 nanowires exhibited the best mechanical properties. Moreover, the Young's modulus was improved for the polymer nanocomposite with MoO_3 nanowires, followed by the mesoporous SiO_2 nanoparticles with an average pore size of 2.6 nm and a total pore volume of 0.17 cm^3/g [44]. The Young's modulus was similar for the composites with the microparticles, where the interactions between the particles and the polymer matrix were almost independent of the particle size and the type of inorganic filler.

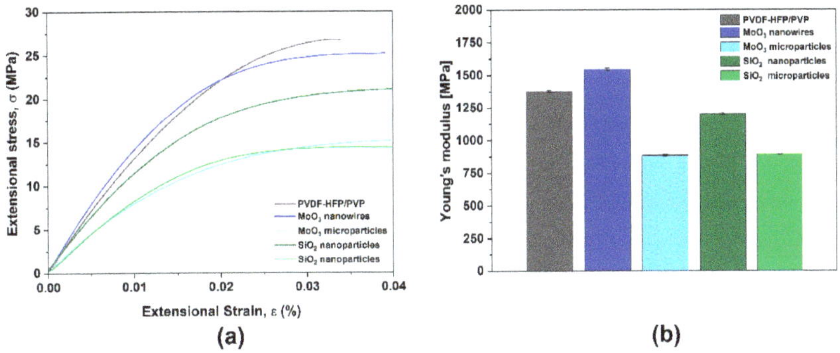

Figure 7. Mechanical properties of the polymer nanocomposites samples obtained with uniaxial mechanical tests. (**a**) Stress-strain curves, (**b**) values of Young's modulus for PVDF-HFP/PVP polymer blend and polymer composites with MoO_3 and SiO_2 nano- and micro-particles.

The Young's modulus for PVDF-HFP is 1560 MPa which decreases to 330 MPa with increasing addition of the $MgCl_2{\cdot}6H_2O$ filler (4 wt. %) [45]. On the other hand, polymer blends consisting of a PVDF/PVP showed a much higher value of Young's modulus (~2930 MPa), which was further improved to ~2960 MPa by the addition of 7 wt.% of lanthanated $CoFe_2O_4$ nanoparticles (CLFO) [46].

The particle size in polymer matrix influences the mechanical behavior of composite due to the different degree of interactions between particles and the host polymer matrix or, in other words, different molecular-chain-network interactions [20]. The particles act as junctions among molecular chains of polymers and additionally improve the interactions in polymer matrix [20].

4. Conclusions

In this study, the influence of incorporating two different types and sizes of particles into a PVDF-HFP/PVP polymer matrix using the solvent casting method to produce thin polymer composite films was investigated. Furthermore, SiO_2 nanoparticles were used as a comparison and possible substitute to the currently-used MoO_3. SiO_2 nanoparticles are widely used, inexpensive and widely available, while the production of MoO_3 nanowires is very limited. Both MoO_3 and SiO_2 particles are promising materials for environmental applications, such as antimicrobial or antifouling coatings.

The presented study showed that the addition of nano/micro MoO_3 or SiO_2 fillers to the polymer matrix, which consists of PVDF-HFP and PVP polymer, improved the thermal and mechanical properties of the polymer composites. In addition, the MoO_3 nanowires formed a geometric network in the polymer matrix, resulting in the best mechanical properties, i.e., the highest elastic modulus, hardness and Young's modulus. This improvement is consistent with the identification of the ferroelectric polar β-crystal phase of the PVDF-HFP polymer in this nanocomposite. On the other hand, MoO_3 microparticles predominantly induced the γ-phase. The SiO_2 particles induced the crystallization of the PVDF-HFP polymer in the γ-phase, but a small amount of α-phase was also identified in the composite with the addition of SiO_2 nanoparticles. However, metal oxide particles in the form of microsized contributed to mechanical properties less effectively due to weaker polymer matrix-filler interactions. The results emphasize the importance of particle size and its distribution in achieving the desired coatings properties.

Author Contributions: Conceptualization, U.G.C.; methodology, M.M. and U.G.C.; validation, U.G.C. and M.M.; formal analysis, U.G.C., M.M., A.P.K. and M.Š.; investigation, U.G.C., M.M., A.P.K. and M.Š.; data curation, M.M.; writing—original draft preparation, U.G.C.; writing—review and editing, M.M., U.G.C. and L.S.P.; visualization, U.G.C. and M.M.; supervision, L.S.P.; project administration, U.G.C.; funding acquisition, L.S.P. All authors have read and agreed to the published version of the manuscript.

Funding: This research was funded by the Slovenian Research Agency, grant number P2-0264 and P1-0099. The APC was funded by the Slovenian Research Agency, P2-0264.

Institutional Review Board Statement: Not applicable.

Informed Consent Statement: Not applicable.

Data Availability Statement: Data are contained within the article.

Acknowledgments: The authors would like to thank Maja Remškar for the synthesis of MoO_3 nanowires.

Conflicts of Interest: The authors declare no conflicts of interest. The funders had no role in the design of the study; in the collection, analyses, or interpretation of data; in the writing of the manuscript; or in the decision to publish the results.

References

1. Ahmed, S.; Masud, A.K.M. Evaluation of Effective Thermal Conductivity of Multiwalled Carbon Nanotube Reinforced Polyme Composites Using Finite Element Method and Continuum Model. *Procedia Eng.* **2014**, *90*, 129–135. [CrossRef]
2. Morampudi, P.; Namala, K.K.; Gajjela, Y.K.; Barath, M.; Prudhvi, G. Review on Glass Fiber Reinforced Polymer Composite *Mater. Today Proc.* **2021**, *43*, 314–319. [CrossRef]
3. Miao, C.; Hamad, W.Y. Cellulose Reinforced Polymer Composites and Nanocomposites: A Critical Review. *Cellulose* **2013**, 2 2221–2262. [CrossRef]
4. Jotiram, G.A.; Palai, B.K.; Bhattacharya, S.; Aravinth, S.; Gnanakumar, G.; Subbiah, R.; Chandrakasu, M. Investigating Mechanica Strength of a Natural Fibre Polymer Composite Using SiO_2 Nano-Filler. *Mater. Today Proc.* **2022**, *56*, 1522–1526. [CrossRef]
5. Vivekchand, S.R.C.; Ramamurty, U.; Rao, C.N.R. Mechanical Properties of Inorganic Nanowire Reinforced Polymer–Matri Composites. *Nanotechnology* **2006**, *17*, S344. [CrossRef]
6. Jordan, J.; Jacob, K.I.; Tannenbaum, R.; Sharaf, M.A.; Jasiuk, I. Experimental Trends in Polymer Nanocomposites—A Revie *Mater. Sci. Eng. A* **2005**, *393*, 1–11. [CrossRef]
7. Cho, J.; Joshi, M.S.; Sun, C.T. Effect of Inclusion Size on Mechanical Properties of Polymeric Composites with Micro and Nan Particles. *Compos. Sci. Technol.* **2006**, *66*, 1941–1952. [CrossRef]
8. Mayeen, A.; Santhosh, A.; Joseph, N.; Jose, J.; Manoj, A.; Joseph, S.; Bhat, S.; John, H. Flexible, Electrospun Boron Nitrid Nanosheets (BNNS)/Hydroxyapatite –PVDF Nanofibers with Superior Piezoelectric/Ferroelectric, Biocompatible Features fo Effective Bone Tissue Regeneration. *J. Alloys Compd.* **2024**, *1002*, 175111. [CrossRef]
9. Kar, E.; Ghosh, P.; Pratihar, S.; Tavakoli, M.; Sen, S. SiO_2 Nanoparticles Incorporated Poly(Vinylidene) Fluoride Composite fo Efficient Piezoelectric Energy Harvesting and Dual-Mode Sensing. *Energy Technol.* **2023**, *11*, 2201143. [CrossRef]
10. Alam, M.M.; Crispin, X. The Past, Present, and Future of Piezoelectric Fluoropolymers: Towards Efficient and Robust Wearabl Nanogenerators. *Nano Res. Energy* **2023**, *2*, e9120076. [CrossRef]
11. Janićijević, A.; Filipović, S.; Sknepnek, A.; Salević-Jelić, A.; Jančić-Heinemann, R.; Petrović, M.; Petronijević, I.; Stamenović M.; Živković, P.; Potkonjak, N.; et al. Structural, Mechanical, and Barrier Properties of the Polyvinylidene Fluoride-Bacteria Nanocellulose-Based Hybrid Composite. *Polymers* **2024**, *16*, 1033. [CrossRef]
12. Al Rashid, A.; Al-Maslamani, N.A.; Abutaha, A.; Hossain, M.; Koç, M. Cobalt Iron Oxide ($CoFe_2O_4$) Reinforced Polyvinyl Alcoho (PVA) Based Magnetoactive Polymer Nanocomposites for Remote Actuation. *Mater. Sci. Eng. B* **2025**, *311*, 117838. [CrossRef]
13. Jiang, Z.; Feng, W.; Lin, Z.; Cai, Y.; Liu, M.; Hu, X.; Zeng, X.; Liu, X.; Wu, Y.; Zhang, Y.; et al. One-Pot Construction of All-in-On Flexible Asymmetrical Supercapacitors for Extreme-Condition Applications. *Chem. Eng. J.* **2024**, *501*, 157553. [CrossRef]
14. Behera, R.; Elanseralathan, K. Modeling of Polyvinylidene Fluoride Nanocomposite Utilizing $BaTiO_3@SiO_2$ for Energy Storage *Mater. Today Proc.* **2022**, *62*, 6554–6560. [CrossRef]
15. Lee, H.; Yamaguchi, K.; Nagaishi, T.; Murai, M.; Kim, M.; Wei, K.; Zhang, K.-Q.; Kim, I.S. Enhancement of Mechanical Propertie of Polymeric Nanofibers by Controlling Crystallization Behavior Using a Simple Freezing/Thawing Process. *RSC Adv.* **2017**, 7 43994–44000. [CrossRef]
16. Chakhchaoui, N.; Farhan, R.; Eddiai, A.; Meddad, M.; Cherkaoui, O.; Mazroui, M.; Boughaleb, Y.; Van Langenhove, L. Im provement of the Electroactive β-Phase Nucleation and Piezoelectric Properties of PVDF-HFP Thin Films Influenced by TiO2 Nanoparticles. *Mater. Today Proc.* **2021**, *39*, 1148–1152. [CrossRef]
17. Zhang, S.; Tong, W.; Wang, J.; Wang, W.; Wang, Z.; Zhang, Y. Modified Sepiolite/PVDF-HFP Composite Film with Enhanced Piezoelectric and Dielectric Properties. *J. Appl. Polym. Sci.* **2020**, *137*, 48412. [CrossRef]

Gradišar Centa, U.; Mihelčič, M.; Bobnar, V.; Remškar, M.; Slemenik Perše, L. The Effect of PVP on Thermal, Mechanical, and Dielectric Properties in PVDF-HFP/PVP Thin Film. *Coatings* **2022**, *12*, 1241. [CrossRef]

Centa, U.G.; Sterniša, M.; Višić, B.; Federl, Ž.; Možina, S.S.; Remškar, M. Novel Nanostructured And Antimicrobial PVDF-HFP/PVP/MoO 3 Composite. *Surf. Innov.* **2020**, *9*, 256–266. [CrossRef]

Jiang, Y.; Tohgo, K.; Yang, H. Study of the Effect of Particle Size on the Effective Modulus of Polymeric Composites on the Basis of the Molecular Chain Network Microstructure. *Comput. Mater. Sci.* **2010**, *49*, 439–443. [CrossRef]

Tcherdyntsev, V.V. Reinforced Polymer Composites. *Polymers* **2021**, *13*, 564. [CrossRef] [PubMed]

Riggleman, R.A.; Toepperwein, G.; Papakonstantopoulos, G.J.; Barrat, J.-L.; de Pablo, J.J. Entanglement Network in Nanoparticle Reinforced Polymers. *J. Chem. Phys.* **2009**, *130*, 244903. [CrossRef]

Woigk, W.; Fuentes, C.A.; Rion, J.; Hegemann, D.; van Vuure, A.W.; Dransfeld, C.; Masania, K. Interface Properties and Their Effect on the Mechanical Performance of Flax Fibre Thermoplastic Composites. *Compos. Part A Appl. Sci. Manuf.* **2019**, *122*, 8–17. [CrossRef]

Žigon, J.; Centa, U.G.; Remškar, M.; Humar, M. Application and Characterization of a Novel PVDF-HFP/PVP Polymer Composite with MoO_3 Nanowires as a Protective Coating for Wood. *Sci. Rep.* **2023**, *13*, 3429. [CrossRef] [PubMed]

Sterniša, M.; Gradišar Centa, U.; Drnovšek, A.; Remškar, M.; Smole Možina, S. Pseudomonas Fragi Biofilm on Stainless Steel (at Low Temperatures) Affects the Survival of Campylobacter Jejuni and Listeria Monocytogenes and Their Control by a Polymer Molybdenum Oxide Nanocomposite Coating. *Int. J. Food Microbiol.* **2023**, *394*, 110159. [CrossRef] [PubMed]

Centa, U.G.; Sterniša, M.; Slemenik, L. *Perše Surface and Mechanical Properties of Polymer Nanocomposite PVDF-HFP/PVP/SiO_2 with Anti-Fouling Properties*; SECTOR: Seville, Spain, 2024; pp. 11–14.

Cao, J.H.; Zhu, B.K.; Xu, Y.Y. Structure and Ionic Conductivity of Porous Polymer Electrolytes Based on PVDF-HFP Copolymer Membranes. *J. Memb. Sci.* **2006**, *281*, 446–453. [CrossRef]

Shir Mohammadi, M.; Hammerquist, C.; Simonsen, J.; Nairn, J.A. The Fracture Toughness of Polymer Cellulose Nanocomposites Using the Essential Work of Fracture Method. *J. Mater. Sci.* **2016**, *51*, 8916–8927. [CrossRef]

Gradišar Centa, U.; Kocbek, P.; Belcarz, A.; Škapin, S.D.; Remškar, M. Polymer Blend Containing MoO_3 Nanowires with Antibacterial Activity against Staphylococcus Epidermidis ATCC 12228. *J. Nanomater.* **2020**, *2020*, 9754024. [CrossRef]

Klapiszewski, Ł.; Zdarta, J.; Szatkowski, T.; Wysokowski, M.; Nowacka, M.; Szwarc-Rzepka, K.; Bartczak, P.; Siwińska-Stefańska, K.; Ehrlich, H.; Jesionowski, T. Silica/Lignosulfonate Hybrid Materials: Preparation and Characterization. *Open Chem.* **2014**, *12*, 719–735. [CrossRef]

Li, Z.; Jiang, X.; Liu, H.; Yao, Z.; Liu, A.; Ming, L. Evaluation of Hydrophilic and Hydrophobic Silica Particles on the Release Kinetics of Essential Oil Pickering Emulsions. *ACS Omega* **2022**, *7*, 8651–8664. [CrossRef]

Dash, S.; Mohanty, H.S.; Ravikant; Kumar, A.; Thomas, R.; Pradhan, D.K. Ferroelectric Ceramic Dispersion to Enhance the β Phase of Polymer for Improving Dielectric and Ferroelectric Properties of the Composites. *Polym. Bull.* **2021**, *78*, 5317–5336. [CrossRef]

Jangra, M.; Gupta, M.; Hussain, S. Probing the Local Interactions and Their Effects on the Dielectric Properties of PVDF/PVP Composite Films. *J. Mol. Struct.* **2024**, *1307*, 137913. [CrossRef]

Kobayashi, M.; Tashiro, K.; Tadokoro, H. Molecular Vibrations of Three Crystal Forms of Poly(Vinylidene Fluoride). *Macromolecules* **1975**, *8*, 158–171. [CrossRef]

Bai, H.; Wang, X.; Zhou, Y.; Zhang, L. Preparation and Characterization of Poly(Vinylidene Fluoride) Composite Membranes Blended with Nano-Crystalline Cellulose. *Prog. Nat. Sci. Mater. Int.* **2012**, *22*, 250–257. [CrossRef]

Cai, X.; Lei, T.; Sun, D.; Lin, L. A Critical Analysis of the α, β and γ Phases in Poly(Vinylidene Fluoride) Using FTIR. *RSC Adv.* **2017**, *7*, 15382 15389. [CrossRef]

7. Borodko, Y.; Habas, S.E.; Koebel, M.; Yang, P.; Frei, H.; Somorjai, G.A. Probing the Interaction of Poly(Vinylpyrrolidone) with Platinum Nanocrystals by UV−Raman and FTIR. *J. Phys. Chem. B* **2006**, *110*, 23052–23059. [CrossRef] [PubMed]

3. Al-Oweini, R.; El-Rassy, H. Synthesis and Characterization by FTIR Spectroscopy of Silica Aerogels Prepared Using Several $Si(OR)_4$ and $R''Si(OR')_3$ Precursors. *J. Mol. Struct.* **2009**, *919*, 140–145. [CrossRef]

9. Mishra, K.; Hashmi, S.A.; Rai, D.K. Protic Ionic Liquid-Based Gel Polymer Electrolyte: Structural and Ion Transport Studies and Its Application in Proton Battery. *J. Solid State Electrochem.* **2014**, *18*, 2255–2266. [CrossRef]

0. Paramee, S.; Guo, R.; Bhalla, A.S.; Manuspiya, H. A Comparison of Shear-mixing and Solvent-induced on Phase Behavior, Thermal and Dielectric Properties of PVDF-HFP/MOF Composites. *J. Appl. Polym. Sci.* **2022**, *139*, e52741. [CrossRef]

1. Lopes, A.C.; Costa, C.M.; Tavares, C.J.; Neves, I.C.; Lanceros-Mendez, S. Nucleation of the Electroactive γ Phase and Enhancement of the Optical Transparency in Low Filler Content Poly(Vinylidene)/Clay Nanocomposites. *J. Phys. Chem. C* **2011**, *115*, 18076–18082. [CrossRef]

2. Lizundia, E.; Reizabal, A.; Costa, C.M.; Maceiras, A.; Lanceros-Méndez, S. Electroactive γ-Phase, Enhanced Thermal and Mechanical Properties and High Ionic Conductivity Response of Poly (Vinylidene Fluoride)/Cellulose Nanocrystal Hybrid Nanocomposites. *Materials* **2020**, *13*, 743. [CrossRef]

3. Yuan, D.; Li, Z.; Thitsartarn, W.; Fan, X.; Sun, J.; Li, H.; He, C. β Phase PVDF-Hfp Induced by Mesoporous SiO_2 Nanorods: Synthesis and Formation Mechanism. *J. Mater. Chem. C Mater.* **2015**, *3*, 3708–3713. [CrossRef]

4. Klapiszewski, Ł.; Królak, M.; Jesionowski, T. Silica Synthesis by the Sol-Gel Method and Its Use in the Preparation of Multifunctional Biocomposites. *Open Chem.* **2014**, *12*, 173–184. [CrossRef]

45. Yuennan, J.; Muensit, N. Preparation and Characterization of Flexible PVDF-HFP Film for Piezoelectric Applications. *IOP C Ser. Mater. Sci. Eng.* **2020**, *715*, 012107. [CrossRef]
46. Bahloul, C.; Ez-Zahraoui, S.; Eddiai, A.; Cherkaoui, O.; Mazraoui, M.; Semlali, F.-Z.; El Achaby, M. Ferrite-Doped Rare-Ea Nanoparticles for Enhanced β-Phase Formation in Electroactive PVDF Nanocomposites. *RSC Adv.* **2024**, *14*, 38872–388 [CrossRef]

coatings

MDPI

rticle

reparation of Chitosan-Coated *Toddalia asiatica* (L.) Lam Extract Microcapsules and Its Effect on Coating Antibacterial Properties

e Zhu [1,2], Ying Wang [1,2] and Xiaoxing Yan [1,2,*]

1 Co-Innovation Center of Efficient Processing and Utilization of Forest Resources, Nanjing Forestry University, Nanjing 210037, China; zhuye@njfu.edu.cn (Y.Z.); wangying1214@njfu.edu.cn (Y.W.)
2 College of Furnishings and Industrial Design, Nanjing Forestry University, Nanjing 210037, China
* Correspondence: yanxiaoxing@njfu.edu.cn

Abstract: Antibacterial microcapsules were prepared using chitosan as the shell material and *Toddalia asiatica* (L.) Lam extract as the core material. The optimal preparation process for the microcapsules of *Toddalia asiatica* (L.) Lam extract were determined via orthogonal and single-factor experiments as follows: the mass ratio of *Toddalia asiatica* (L.) Lam extracts and chitosan (M_T:M_C) was 3.0:1, the pH value of microencapsulation was 7, and the reaction temperature was 50 °C. The M_T:M_C increased, the glossiness of the coatings increased and then decreased, the color difference and roughness of the coatings showed an increasing trend, the transmittance of the coatings decreased, the grade of citric acid resistance decreased, the resistance to ethanol and detergents of the coatings increased, and the fracture elongation of the coating increased and then fell. As the M_T:M_C of microcapsules increased, the antibacterial rates of the coating against *Escherichia coli* increased first, then decreased, and lastly increased. The trend of the coating against *Staphylococcus aureus* first increased and then decreased. When the microcapsule M_T:M_C was 3.0:1, the performance of the coating was better, and the antibacterial rates against the two bacteria were 67.14% and 68.39%, respectively. *Toddalia asiatica* (L.) Lam extracts endow waterborne coatings with antibacterial properties, expanding the application range of waterborne coatings.

Keywords: *Toddalia asiatica* (L.) Lam extracts; microcapsules; waterborne coatings; antibacterial coatings

1. Introduction

With the improvement of the economy and living standards, health and hygiene issues have received increasing attention from people [1]. However, the pollution caused by microorganisms in the environment where people live is extremely severe [2]. The use of antibacterial materials is the simplest and most effective way to achieve a pollution-free environment [3–5]. On the premise of not changing the original adhesion of the coating, an ideal antibacterial coating should meet the characteristics of low cost, low toxicity, high antibacterial efficiency, broad-spectrum antibacterial properties, and antibacterial persistence [6–8]. The antibacterial agents added to antibacterial coatings are mainly divided into three types: natural antibacterial agents [9], organic antibacterial agents [10], and inorganic antibacterial agents [11]. Natural antibacterial agents are non-toxic, have a wide range of sources, and have significant antibacterial effects, but their stability is poor [12,13]. Natural antibacterial agents are prepared into microcapsules to improve their stability and corrosion resistance [14]. These antibacterial microcapsules are added to waterborne coatings to endow them with antibacterial properties, which is of great practical significance for improving the service life of coatings, enriching the functions of waterborne coatings, and protecting substrates [15–17].

With the proposal of the "dual carbon" goal, while developing antibacterial and antiviral coatings, the toxicological properties of coatings have also been given attention [18].

The production concept of green, non-toxic, and sustainable development is advocated [19–21]. *Toddalia asiatica* (L.) Lam extracts belong to the Rutaceae family, and its parts have extremely high medicinal value [22]. The chemical components of its anhydrous ethanol extract have good antibacterial properties against *Staphylococcus aureus* and *Escherichia coli* [23]. He et al. isolated resveratrol from the *Toddalia asiatica* (L.) Lam and tested it against *Gram-positive bacteria, methicillin-resistant Staphylococcus aureus*, and *broad-spectrum antibiotics β- Lactamase Staphylococcus aureus* with it having strong antibacterial activity, and the minimum inhibitory concentration against all three bacteria was 0.156 mg/mL [24]. Duraipandiyan et al. [25] used hexane, chloroform, ethyl acetate, methanol and water extracts of *Toddalia asiatica* leaves and isolated the compound Flindersine to test against bacteria and fungi. They found that ethyl acetate extract showed promising antibacterial and antifungal activity and the isolated compound Flindersine showed moderate activity against bacteria and fungi. The existing commercial antibacterial products mainly include antibacterial additives such as Miao Kangbao, Biocote, and SteriTouchden, as well as antibacterial coatings such as SilverSan, InterponAM, and Nippon Cleansing 120 bamboo charcoal antibacterial [26]. The main reason why antibacterial coatings can exert antibacterial effects is that special antibacterial agents are added to the coating formula, which can have a sustained antibacterial ability after coating is completed [27]. Chitosan was selected as the shell material and *Toddalia asiatica* (L.) Lam extracts were selected as the core material for the preparation of microcapsules. They are both natural products that are environmentally friendly. At the same time, microcapsule technology was used to overcome the weakness of the poor stability of natural antibacterial agents. The shell material chitosan also has antibacterial properties, and it was synergized with the core material antibacterial agent, effectively improving the antibacterial performance against *Staphylococcus aureus* and *Escherichia coli*. This coating had more significant sustainable properties compared to other similar technologies on the market due to its more certain composition, ease of addition, mixing and dispersion, and stable mixing.

Chitosan was selected as the shell material and *Toddalia asiatica* (L.) Lam extracts was selected as the core material for the preparation of microcapsules. A three factor, two level orthogonal experiment was designed to explore the optimal preparation process of microcapsules by analyzing factors such as yield, encapsulation rate, microstructure, and chemical composition. On this basis, a single-factor experiment was conducted. The obtained single-factor microcapsules were added to the waterborne topcoat at the same content to prepare an antibacterial coating, and the effect of the mass ratio of chitosan and *Toddalia asiatica* (L.) Lam extracts on the antibacterial coating was investigated. In this study, *Toddalia asiatica* (L.) Lam extracts were prepared into microcapsules and added to the coating. This not only improved the poor stability of natural antibacterial agents, but also enhanced the antibacterial efficacy of the coating, reduced environmental pollution, and maintained the stability of the substrate itself, expanding the application range of waterborne coatings.

2. Materials and Methods

2.1. Materials

Table 1 shows the list of experimental materials. The *Toddalia asiatica* (L.) Lam Leaves were from Lingshan County, Qinzhou City, Guangxi Province, China. The leaves were placed in a 40 °C oven (Shanghai Jinghong Experimental Equipment Co., Ltd., Shanghai, China) and dried to a constant weight. They were ground into powder using a crusher (Zhejiang Fengli Crushing Equipment Co., Ltd., Shengzhou, China). The coating was a waterborne acrylic topcoat (Jiangsu Haitian Technology Co., Ltd., Wuxi, China). A 50 mm × 50 mm × 10 mm sized silicone mold was selected to prepare the coating. The size of the glass substrate (Jiangsu Huaou Glass Co., Ltd., Yancheng, China) was 25 mm × 75 mm × 10 mm.

Table 1. List of experimental materials.

Test Material	Purity	Manufacturer
Chitosan	AR	Shandong Haiyi Chemical Technology Co., Ltd., Binzhou, China
Acetic acid	AR	Henan Mingao Chemical Co., Ltd., Anyang, China
Tripolyphosphate (TPP)	AR	Sichuan Blue Sword Chemical Group Co., Ltd., Shifang, China
Tween-80	AR	Jinan Huifengda Chemical Co., Ltd., Jinan, China
X-100	AR	Shanghai Maokang Biotechnology Co., Ltd., Shanghai, China
Anhydrous ethanol	AR	Jinan Hongrun Chemical Co., Ltd., Jinan, China
Nutrient agar medium	-	Huankai Microbial Technology Co., Ltd., Guangzhou, China
Nutrient broth	-	Qingdao Haibo Biotechnology Co., Ltd., Qingdao, China
Staphylococcus aureus	-	Ningbo Mingzhou Biotechnology Co., Ltd., Ningbo, China
Escherichia coli	-	Ningbo Mingzhou Biotechnology Co., Ltd., Ningbo, China
Cleaning agents	AR	Guangdong Baiyun Cleaning Group Co., Ltd., Guangzhou, China
Citric acid	AR	Shandong Lemon Biochemical Co., Ltd., Weifang, China

2.2. *Preparation Method and Experimental Design of Microcapsules*

2.2.1. Preparation Method of Microcapsules

First, 0.20 g of acetic acid was diluted in 19.8 g of deionized water in a beaker (Ningbo Pulai Experimental Instrument Co., Ltd., Ningbo, China) and a 1% concentration acetic acid solution was prepared. A 0.20 g of chitosan powder was weighed and added to the acetic acid solution and then a magnetic stirrer (Yancheng Xieying Machinery Manufacturing Co., Ltd., Yancheng, China) was added. The beaker was placed in a water bath pot (Gongyi Yuhua Instrument Co., Ltd., Gongyi, China) with a set speed of 600 rpm and a temperature of 60 °C until the chitosan powder was completely dissolved to form a chitosan shell material solution. The leaf powder was mixed with anhydrous ethanol in a mass ratio of 1:15 in the beaker, and the beaker was sealed with cling film. The concentration of blade powder was 6.25%. The beaker was placed in the water bath pot for extraction for 70 min with a set speed of 600 rpm and a temperature of 60 °C. The extracted mixture was put into a centrifuge (Shanghai Luxiangyi Centrifuge Instrument Co., Ltd., Shanghai, China) and centrifuged for 7 min at a speed of 7000 rpm. The mixture was filtered using a vacuum pump (Ningbo Aifake Vacuum Technology Co., Ltd., Ningbo, China) combined with a Buchner funnel (Wuhan Dingsheng Zhongtian Experimental Instrument Co., Ltd., Wuhan, China) to obtain the filtrate. The obtained filtrate was subjected to rotary evaporation and dried to obtain a solid extract, sealed, and then stored in a cool and dark place for future use [28]. Tween-80 and X-100 were weighed to prepare an emulsifier solution with a mass fraction of about 3.0%. The solution was added dropwise to the *Toddalia asiatica* (L.) Lam extracts. The speed of the water bath pot was set to 600 rpm and the temperature was 50 °C. The solution was placed in the water bath pot for 45 min. The prepolymer solution was sucked up by a dropper and added to the core material lotion drop by drop. The core material (0.20 g) and shell material (0.40–0.80 g) solution were fully mixed for 1 h. The concentration range of the core material was 0.25–0.45%, and the concentration range of the shell material was 0.89–1.01%. The 0.5 mol/mL NaOH solution was added to adjust the pH value of the solution to around 7. The TPP solution with a concentration of 5.0% was added; then, the beaker was placed in the water bath for the crosslinking reaction and microencapsulated for 3 h. The speed of the water bath was set to 600 rpm and the temperature was 60 °C. The obtained solution was cooled at room temperature. The high-speed centrifuge speed was set to 7000 rpm, and the resulting product was centrifuged for 10 min. The upper clear liquid was poured out, and the solid product was placed in a freeze-drying machine (Ningbo Xinzhi Freeze Drying Equipment Co., Ltd., Ningbo, China) for 24 h before being taken out and ground. The final product obtained was a powdered microcapsule [29].

2.2.2. Orthogonal and Single-Factor Experimental Design of Microcapsules

According to the preliminary experimental results, the main influencing factors in the preparation process of chitosan-coated *Toddalia asiatica* (L.) Lam extract microcapsules included M_T:M_C, the pH value of the solution during the crosslinking reaction, and the reaction temperature. An orthogonal experiment was designed to investigate the effects of M_T:M_C, the pH value of the solution during the crosslinking reaction, and the reaction temperature on the preparation of microcapsules. A single-factor experiment was designed to investigate the effect of M_T:M_C on the preparation of microcapsules, based on the optimal preparation parameters obtained by combining the yield and encapsulation rate of the orthogonal experiments. Table 2 shows the material table for the orthogonal experiments. Table 3 shows the material table for single-factor experiments.

Table 2. Orthogonal test bill of materials.

Sample (#)	Chitosan (g)	*Toddalia Asiatica* (L.) Lam Extracts (g)	Ethanol (g)	Tween-80 (g)	X-100 (g)	Deionized Water (g)	TPP (g)	Deionized Water (g)
1	0.20	0.40	7.60	0.07	0.73	25.87	0.50	9.50
2	0.20	0.40	7.60	0.07	0.73	25.87	0.50	9.50
3	0.20	0.60	10.80	0.10	1.10	38.80	0.50	9.50
4	0.20	0.60	10.80	0.10	1.10	38.80	0.50	9.50

Table 3. Single-factor test bill of materials.

Sample (#)	Chitosan (g)	*Toddalia Asiatica* (L.) Lam Extracts (g)	Ethanol (g)	Tween-80 (g)	X-100 (g)	Deionized Water (g)	TPP (g)	Deionized Water (g)
5	0.20	0.40	7.60	0.07	0.73	25.87	0.50	9.50
6	0.20	0.50	9.50	0.08	0.92	32.33	0.50	9.50
7	0.20	0.60	11.40	0.10	1.10	38.80	0.50	9.50
8	0.20	0.70	13.30	0.12	1.28	45.27	0.50	9.50
9	0.20	0.80	15.20	0.13	1.47	51.73	0.50	9.50

2.3. Preparation Method of the Coating

The prepared microcapsules were added to the waterborne topcoat with a mass fraction of 5% and mixed evenly to obtain 1.0 g of topcoat coating. The obtained topcoat coating was evenly applied in the silicone mold, leveled, and dried at room temperature. The coating was dried in the oven at a set temperature of 55 °C for 30 min.

2.4. Testing and Characterization

2.4.1. Performance Characterization of Microcapsules

(1) Coverage rate (C): The microcapsules were soaked in ethanol for 24 h and then filtered and dried. The formula given calculates the coverage rate, quantifying the extent of the shell material covering the microcapsule surface (1). M_1 is the weight of microcapsules, M_2 is the weight of the filter paper, and M_3 is the weight of the total mass of the dried filter paper and shell material.

$$C = \frac{(M_1 + M_2) - M_3}{M_1} \quad (1)$$

(2) Yield rate (Y): The total mass, M_1, includes all materials. After drying, the microcapsule powder mass is M_2. The yield rate calculation is shown in the Formula (2).

$$Y = \frac{M_2}{M_1} \quad (2)$$

(3) Microscopic morphology and chemical composition: Microcapsule morphology was observed via a Zeiss optical microscope (OM, Carl Zeiss AG, Oberkochen, Germany). Scanning electron microscopy (SEM, Tescan, Brno, the Czech Republic) was employed to scrutinize the microcapsule and coating microstructures. Fourier-transform infrared spectroscopy (FTIR, Brucker AG, Karlsruhe, Germany) was used to determine the chemical composition of both the microcapsules and the coatings.

2.4.2. Color Difference Testing of the Coating

In the light of GB/T 11186.3-1989 [30], a SEGT-J colorimeter (Zhuhai Tianchuang Instrument Co., Ltd., Zhuhai, China) was utilized to assess and record the chromaticity of the coating. The color difference (ΔE) between the microcapsule-containing coating and the pure coating was determined using Formula (3), taking into account ΔL (brightness variation), Δa (red–green difference), and Δb (yellow–blue difference).

$$\Delta E = \left[(\Delta L)^2 + (\Delta a)^2 + (\Delta b)^2\right]^{\frac{1}{2}} \tag{3}$$

2.4.3. Glossiness, Transmittance, and Light Loss Testing of the Coating

In GB/T 4893.6-2013 [31], a glossmeter (Shenzhen Linshang Technology Co., Ltd., Shenzhen, China) was used to test and record the glossiness values of the coating at three incidence angles of 20°, 60°, and 85°, with the unit being GU.

A UV spectrophotometer (Beijing Puxi General Instrument Co., Ltd., Beijing, China) was employed to determine and analyze the transmittance of the coating across the visible light spectrum, ranging from 380 to 780 nm.

The light loss incurred by the coating was subsequently quantified using Formula (4), where G_L denotes the light loss rate, G_0 represents the glossiness of the coating in its unmodified state (without microcapsules), and G_1 signifies the glossiness after the incorporation of microcapsules.

$$G_L = \frac{G_0 - G_1}{G_0} \times 100\% \tag{4}$$

2.4.4. Roughness and Tensile Testing of the Coating

The roughness value was tested and recorded using a JB-4C roughness tester (Cangzhou Oupu Testing Instrument Co., Ltd., Cangzhou, China).

Silicone molds were used for curing and demolding of the coatings. A universal mechanical testing machine (Instron, Boston, MA, USA) was used to conduct tensile tests on the coatings. The fracture elongation of the coating, calculated using Formula (5), quantifies the elongation (e) at the fracture point, taking into account the original length (L_1) and the length at fracture (L), providing a measure of the coating's tensile properties.

$$e = \frac{L - L_1}{L_1} \times 100\% \tag{5}$$

2.4.5. Cold Liquid Resistance

According to GB/T 4893.1-2021 [32], a citric acid solution with a mass fraction of 10%, undescended ethanol with a volume fraction of 96%, and a cleaning agent prepared with deionized water were chosen as the experimental liquids. Damage to the surface was inspected under specified lighting conditions. The test results were evaluated using numerical levels.

2.4.6. Antibacterial Properties of the Coating

Adhering to GB/T 21866-2008 [33], the testing protocol was meticulously executed. *Escherichia coli* (ATCC25922) and *Staphylococcus aureus* (ACTT6538) were selected for antibacterial testing. An agar plate medium was prepared by accurately weighing 24 g of agar medium into 1000 mL of distilled water, followed by sterilization. The slant-preserved bacterial strains were then inoculated onto the agar plates and incubated in a constant temperature and humidity incubator (Shanghai Zhetu Scientific Instrument Co., Ltd., Shanghai, China) at 38 °C for 18–20 h. The 9 g of nutrient broth and 500 mL of distilled water were weighed to prepare the nutrient broth culture medium. The polyethylene film was soaked in 70.0% ethanol solution for 30 min, washed with eluent, dried, and set aside for later use. Fresh bacteria from the agar medium were transferred into this broth using an inoculation ring. The sterilized nutrient broth culture medium was sequentially diluted in a ten-fold increase to form a 1:1000 bacterial suspension. Approximately 0.5 mL of this bacterial suspension was then dispensed onto both the test coating and the control group surfaces. A sterilized plastic film was applied to each specimen's surface using sterilized tweezers, and the specimens were placed in sterilized culture dishes before being transferred to a constant temperature and humidity incubator set at 38 °C and RH > 90% for 24 h of cultivation. Following the incubation period, the samples were removed, and 20 mL of eluent was added. The samples were thoroughly rinsed by clamping the thin film with tweezers and repeatedly flushing it. The rinse solution was then inoculated into nutrient agar culture medium and re-incubated in the same incubator conditions for an additional 48 h. Finally, in accordance with GB/T 4789.2-2022 [34], the colony count in the culture medium was meticulously measured and recorded using a colony counter (Guansen Biotechnology (Shanghai) Co., Ltd., Shanghai, China). After 48 h of cultivation, the actual number of colonies recovered from each sample was scaled up by a factor of 1000 for easier counting or standardization purposes. The formula utilized to calculate the antibacterial rate, as presented in Formula (6), designates R as the antibacterial rate, measured in percentage (%). In this formula, B represents the average number of colonies recovered from pure coating samples after 48 h, expressed in CFU/piece. Similarly, C denotes the average number of bacteria recovered from the antibacterial coating sample after the same duration, expressed in CFU/piece.

$$R = \frac{B - C}{B} \times 100\% \quad (6)$$

3. Results and Discussion

3.1. Analysis of Microcapsule Preparation Results

3.1.1. Analysis of the Yield and Encapsulation Rate Results of the Microcapsules

Table 4 shows the analysis of the microcapsule yield results obtained from the orthogonal experiments. In the four groups of microcapsule samples in the orthogonal experiment, the difference in yield was very small, and the yield difference of the four samples fluctuated around 0.5%. From the range of results, the reaction temperature had the greatest impact on the yield of microcapsules. For the yield of microcapsules, the primary and secondary factors affecting their preparation process were C > A > B, and the optimal preparation process was A1, B2, and C2. Table 5 shows the analysis of variance results for yield. The variance results for the three factors were the same as the range results, and none of the three influencing factors were significant.

Table 6 shows the analysis of the microcapsule encapsulation rates obtained from the orthogonal experiments. The coverage rate is an important factor used to evaluate the encapsulation effect of microcapsules and affects their antibacterial properties. Sample 3# had the highest encapsulation rate of 29.71%. The minimum coverage rate was sample 2#, and the coverage rate result was 20.20%. Based on their range results, the primary and secondary levels of three factors were A > B > C. The factor A (M_T:M_C) had the greatest impact on the encapsulation rate of microcapsules. Based on the variance results in Table 7, the optimal preparation process was A2, B1, and C2. The optimal preparation process parameters for the microcapsules were M_T:M_C of 3.0:1, a microencapsulation pH value of 7, and a reaction temperature of 50 °C.

Table 4. Analysis of the yield of the microcapsules during the orthogonal test.

Sample (#)	Factor A M_T:M_C	Factor B The pH of the Reaction	Factor C Reaction Temperature (°C)	Yield Rate (%)
1	2.0:1	7	50	13.78
2	2.0:1	9	60	14.15
3	3.0:1	7	60	13.84
4	3.0:1	9	50	13.76
Mean 1	13.965	13.810	13.770	
Mean 2	13.800	13.955	13.995	
Range	0.165	0.145	0.225	
Factor primary and secondary levels		C > A > B		
Optimal solution		A1 B2 C2		

Table 5. Variance analysis table of yield results.

Factor	Sum of Squared Deviations	Degree of Freedom	F Ratio	F Critical Value	Significance
A	0.027	1	0.818	10.100	
B	0.021	1	0.636	10.100	
C	0.051	1	1.545	10.100	
Error	0.10	3			

Table 6. Analysis on the results of the orthogonal test of the microcapsule coverage rate.

Sample (#)	Factor A M_T:M_C	Factor B The pH of the Reaction	Factor C Reaction Temperature (°C)	Coverage Rate (%)
1	2.0:1	7	50	22.16
2	2.0:1	9	60	20.20
3	3.0:1	7	60	29.71
4	3.0:1	9	50	25.88
Mean 1	21.180	25.935	24.020	
Mean 2	27.795	23.040	24.955	
Range	5.860	2.895	0.935	
Factor primary and secondary levels				A > B > C
Optimal solution				A2 B1 C2

Table 7. Analysis of variance table for coverage rate.

Factor	Sum of Squared Deviations	Degree of Freedom	F Ratio	F Critical Value	Significance
A	43.758	1	2.476	10.100	
B	8.381	1	0.474	10.100	
C	0.874	1	0.049	10.100	
Error	53.01	3			

A single-factor experiment was designed and conducted based on the optimal preparation parameters obtained from the yield and coverage rate of orthogonal experiments. Using M_T:M_C, which had a relatively large influencing factor, as the experimental variable, the pH value of the solution during fixed microencapsulation was 7, and the reaction temperature was 50 °C (determined based on the results of coverage rate) for single-factor testing. Table 8 shows the yield and coverage rate results of the single-factor experimental microcapsules. The highest yield of sample 5# was 30.58%, while the lowest yield of sample 9# was 22.97%. The yield of the microcapsules decreased with the increase in M_T:M_C. This is because, under the premise of maintaining the quality of the shell material, as the quality of the *Toddalia asiatica* (L.) Lam extracts increase, there will be an excess of core material that is not covered. Sample 7# had the highest coverage rate of 37.74%, while Sample 5# had the lowest coverage rate of 28.73%. The coverage rate of microcapsules showed a trend of first increasing and then decreasing with the increase in M_T:M_C. This is because as the quality of the core material increases, the concentration of emulsifiers remains unchanged. When the core material is excessive, the emulsification effect on the core material is poor and not fully emulsified, resulting in a decrease in the coverage rate of the microcapsules.

Table 8. The yield and coating rate of the microcapsules were tested via a single-factor experiment.

Sample (#)	MT:MC	Yield Rate (%)	Coverage Rate (%)
5	2.0:1	30.58	28.73
6	2.5:1	30.27	30.31
7	3.0:1	28.23	37.74
8	3.5:1	27.35	36.54
9	4.0:1	22.97	36.36

3.1.2. Microscopic Morphology Analysis of the Microcapsules

Figure 1 shows the SEM images of five microcapsule samples prepared in a single-factor experiment. When M_T:M_C was 2.0:1, there were fewer successfully prepared microcapsules and more irregular aggregates present. When M_T:M_C was 2.5:1, many spherical microcapsules of different sizes were observed, but there were still many ir-regular aggregates present. When M_T:M_C was 3.0:1, the prepared microcapsules exhibited adhesion and uneven size. When M_T:M_C was 3.5:1 and 4.0:1, there were many mi-crocapsule microspheres, but the microcapsules exhibited uneven dispersion and local adhesion. This is because the core material content is too high and not completely emulsified and dispersed, resulting in an increase in the viscosity of the core material emulsion. During the process of microencapsulation, the core material will be present between the encapsulated microcapsules, leading to the formation of adhesion between them, and resulting in a slightly sticky and rough texture of the formed microcapsule powder [35].

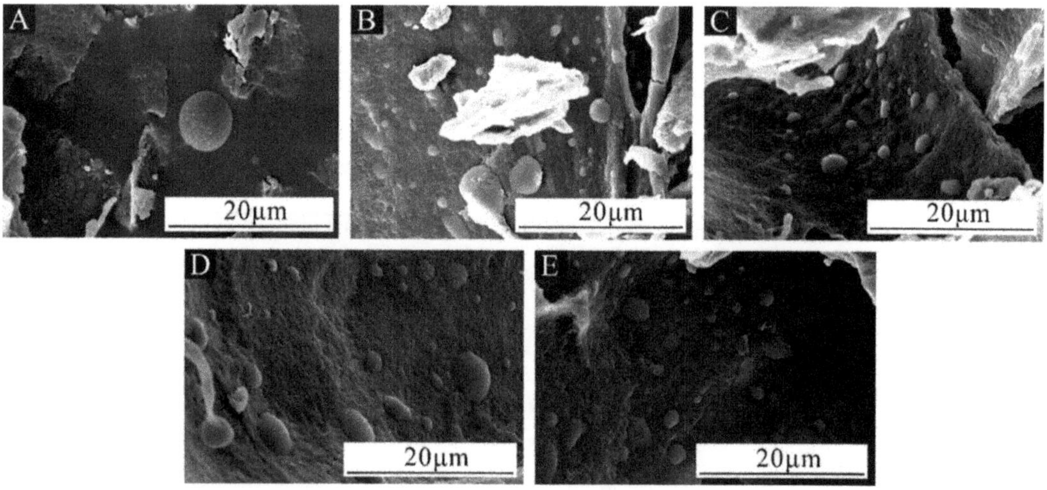

Figure 1. SEM images of microcapsules under the single-factor test: (**A**) Sample 5#, (**B**) Sample 6#, (**C**) Sample 7#, (**D**) Sample 8#, and (**E**) Sample 9#.

As shown in Figure 2, the particle size of the 5# microcapsule sample was widely distributed between 3.0 μm and 5.0 μm. The particle size of the 6# microcapsule sample was concentrated between 2.0 μm and 4.0 μm. The particle size of 7# microcapsules was mainly distributed in 3.0–5.0 μm. The particle size of 8# microcapsules was concentrated between 2.0 μm and 7.0 μm. The particle size distribution of microcapsules was relatively uniform compared to the first four groups of samples, with a distribution between 2.0 μm and 12.0 μm. The particle size distribution of microcapsule samples 8# and 9# was relatively uniform.

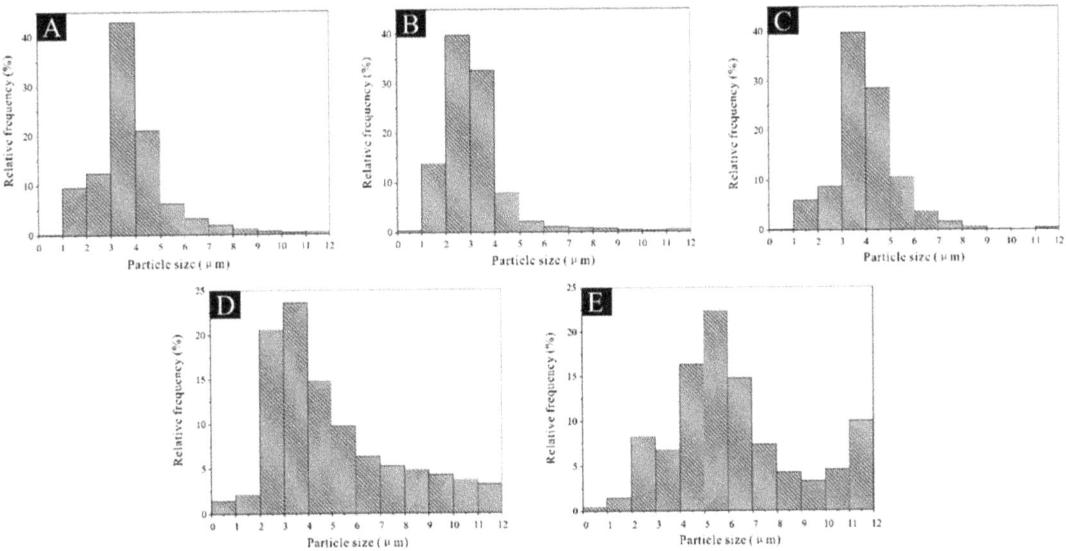

Figure 2. Particle size distribution: **(A)** 5#, **(B)** 6#, **(C)** 7#, **(D)** 8#, and **(E)** 9#.

3.1.3. Chemical Composition Analysis of the Microcapsules

As shown in Figure 3, the absorption peak at 3350 cm^{-1} was the stretching vibration peak of C–O in the core material. The characteristic peak at 1655 cm^{-1} was the stretching absorption peak of C=O in chitosan. The characteristic peak at 1590 cm^{-1} was the deformation vibration peak of $-NH_2$ in chitosan [36]. The two stretching vibration absorption peaks generated at 2921 cm^{-1} and 2875 cm^{-1} were the absorption peaks of C–H in chitosan, and the absorption peak of –OH in chitosan was at 3432 cm^{-1}, indicating the presence of shell material chitosan in the microcapsules [37]. The absorption peaks at 1600 cm^{-1} and 1110 cm^{-1} were characteristic peaks of C=N and C–O in coumarin compounds in the core material, which are positioned on the absorption curve of the microcapsules, proving the presence of the core material *Toddalia asiatica* (L.) Lam extract in the microcapsules as well. The above proves that microcapsules had been successfully prepared.

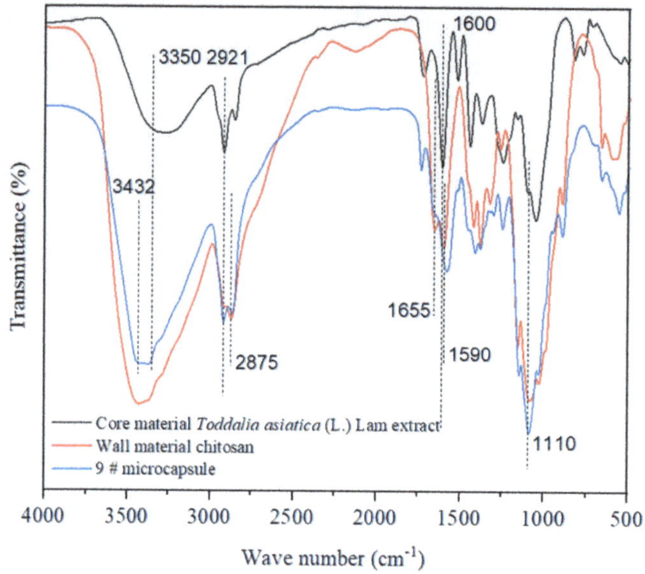

Figure 3. Infrared spectra of core materials, shell materials, and microcapsules.

3.2. Microscopic Morphology and Chemical Composition Analysis of the Coating

Figure 4 shows SEM images of the coating prepared using microcapsules with different M_T:M_C at a 5.0% additive content. When the microcapsule content in the coating was 0%, the surface was relatively flat and smooth. The surface of the antibacterial coating with the addition of 5.0% microcapsules becomes rough, with many protrusions. The surface of the antibacterial coating with samples 5# and 6# added was rougher because there are more irregular aggregates in these two microcapsules, resulting in poor surface smoothness of the prepared coating. The coating of samples 7#, 8#, and 9# had slightly lower roughness compared to the antibacterial coatings of samples 5# and 6#. However, due to the uneven dispersion and local adhesion of these three microcapsule samples, the surface of the prepared coatings was rougher and more uneven than that of the pure coating.

As shown in Figure 5, the absorption peak at 1729 cm^{-1} was the absorption peak of C=O in waterborne coatings [38]. There were the stretching vibration peaks of N–H and O–H at 3425 cm^{-1} and the two stretching vibration absorption peaks of C–H in shell material chitosan appeared at 2921 cm^{-1} and 2875 cm^{-1}, respectively [39]. The characteristic peak appearing at 1114 cm^{-1} was the absorption peak of C–O in the microcapsule core material. This proved that after adding the prepared microcapsules to the waterborne topcoat, the shell and core components belonging to the microcapsules still existed, and there was no chemical reaction between the microcapsules and the waterborne topcoat.

Figure 4. *Cont.*

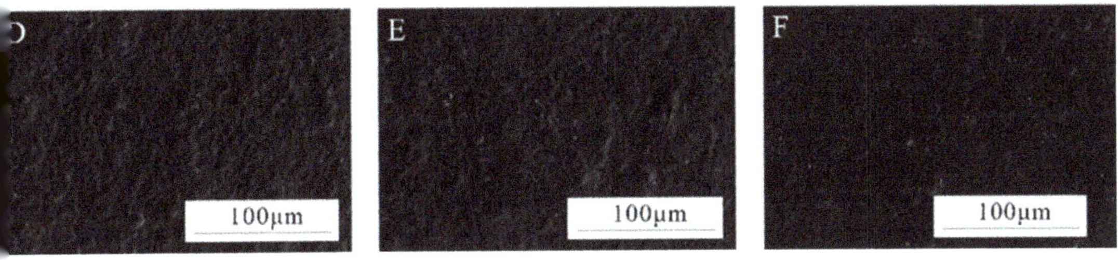

Figure 4. SEM images of coatings prepared by adding 5.0% microcapsules with different M_T:M_C: (**A**) coating without microcapsules, (**B**) coating with microcapsules 5#, (**C**) coating with microcapsules 6#, (**D**) coating with microcapsules 7#, (**E**) coating with microcapsules 8#, and (**F**) coating with microcapsules 9#.

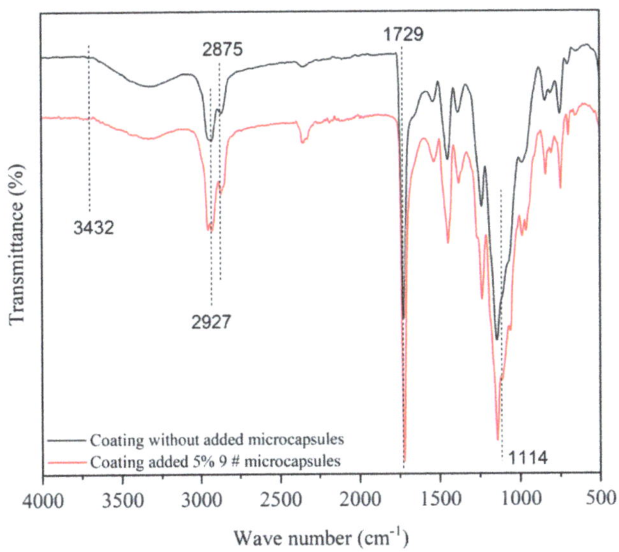

Figure 5. Infrared spectra of the coating.

3.3. *Influence of Different M_T:M_C of Microcapsules on the Optical Properties of the Coatings*

Glossiness is an important indicator for the surface coating of wooden products. Table 9 shows the effect of microcapsules with different M_T:M_C on the glossiness and light loss of coatings. 0# represents the glossiness value of the coating without microcapsules added. Figure 6A shows the changes in the surface glossiness and light loss of the coating measured at three incident angles. Compared with the coating without microcapsules, the addition of microcapsules greatly reduced the glossiness of the coating itself. The difference in glossiness between coatings varied relatively little at an incidence angle of 60°. As the M_T:M_C increased, the glossiness showed a trend of first increasing and then decreasing. The maximum value reached 7.47 GU when the M_T:M_C was 2.5:1, followed by 7.33 GU; at this time, the M_T:M_C was 3.0:1. As shown in Figure 6B, as the M_T:M_C increased, the light loss of the coating showed a trend of first decreasing and then increasing. The light loss of the coating at a 60° incidence angle reached the minimum value of 59.69% when the M_T:M_C was 2.5:1, followed by 60.44%; at this time, the M_T:M_C was 3.0:1. This is because the prepared microcapsule powder is relatively sticky and rough. Adding it to the waterborne coating for curing reduces the smoothness of the coating, thereby reducing the surface light reflection performance of the coating [40,41]. When the added microcapsules M_T:M_C was

2.5:1 and 3.0:1, the glossiness and light loss of the coating were relatively excellent, and the coating performance was better.

Table 9. Effect of microcapsules with different M_T:M_C on the glossiness of the coatings.

Sample (#)	Surface Glossiness of the Coating (GU)			Light Loss (%)		
	20°	60°	85°	20°	60°	85°
0	5.97	18.53	39.57	-	-	-
5	1.67	7.17	4.70	72.03	61.31	88.12
6	1.73	7.47	7.13	71.02	59.69	81.98
7	1.53	7.33	9.80	74.37	60.44	75.23
8	1.43	6.77	8.90	76.05	63.46	77.51
9	1.37	6.67	9.47	77.05	64.00	76.07

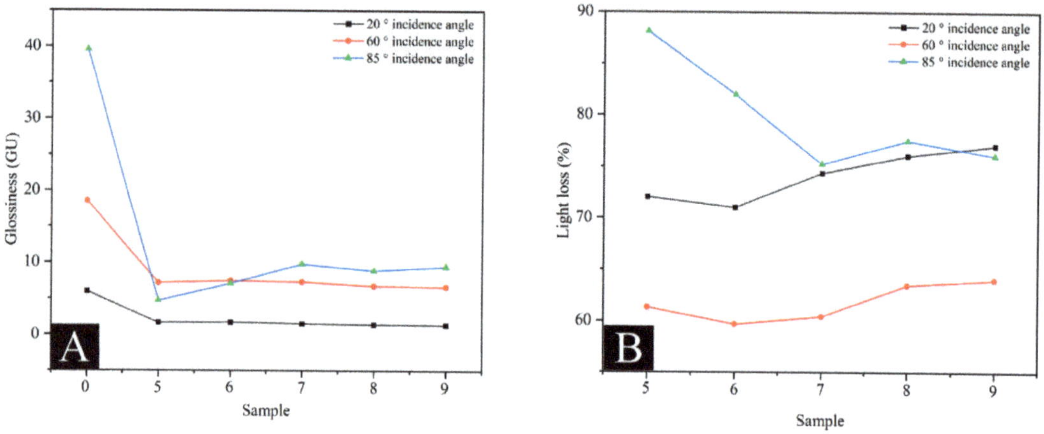

Figure 6. Effect of M_T:M_C of microcapsules on (**A**) the glossiness and (**B**) light loss of the coating.

Table 10 shows the changes in chromaticity and the color difference values of coatings with different M_T:M_C. As shown in Figure 7A, the brightness value of the coating showed a fluctuating downward trend with the increase in M_T:M_C. The addition of microcapsules reduced the reflection ability of the coating to light, thereby reducing the brightness value of the coating. Figure 7B,C show an overall upward trend in the red green and yellow blue values of the coating. This is because the microcapsules are brown in color, and as the core material amount increases, the color of the microcapsules gradually darkens, resulting in an increase in the red–green and yellow–blue values of the coating. As shown in Figure 7D, the color difference value of the coating generally showed an upward trend with the increase in M_T:M_C. When the M_T:M_C was 2.5:1, the minimum color difference value was 18.29. When the M_T:M_C was 4.0:1, the maximum color difference value was 27.17. This is because microcapsules themselves are relatively rough and have a strong sense of particle size. After adding the coating, there were uneven dispersion, local agglomeration, and other phenomena, which increased the color difference of the coating and affected its aesthetics [42,43].

Figure 8 shows the variation trend of transmittance of coatings prepared using microcapsules with different M_T:M_C. Compared with the coating without microcapsules, the overall transmittance of the coating-added microcapsules was lower. As the M_T:M_C continues to increase, the overall transmittance of the prepared coating in the visible light band shows a trend of first increasing and then decreasing. Table 11 shows the transmittance of waterborne coatings containing different M_T:M_C microcapsules in the visible light band. When the added microcapsule M_T:M_C was 2.5:1 (sample 6#) and 3.0:1 (sample 7#), the transparency of the coating was relatively excellent, with corresponding visible light transmittance of 87.05% and 86.00%, respectively. When adding M_T:M_C at 4.0:1 (sample 9#), the light transmittance of the coating was poor, corresponding to a visible light transmittance of 84.08%. Due to the increasing aggregation phenomenon of microcapsules, the surface roughness of the prepared coating is enhanced, which reduces the transmission and reflection of incident light, enhances the scattering phenomenon of light, and thus reduces the transmittance of the coating. When the M_T:M_C was 4.0:1, the minimum transmittance appeared. Because the coating containing the 9# microcapsule has a darker color, this reduces the penetration ability of light and the transmittance of the coating.

Table 10. Effect of microcapsules with different M_T:M_C on the chromaticity and color difference of the coating.

Sample (#)	Chromaticity Parameter						ΔE
	L_1	a_1	b_1	L_2	a_2	b_2	
0	84.30	1.35	5.41	84.65	1.25	5.70	-
5	68.75	7.00	14.25	68.8	6.80	14.05	18.76
6	68.70	7.75	12.50	68.65	7.60	11.85	18.29
7	65.45	10.25	16.85	65.20	9.55	15.45	23.53
8	66.20	10.20	16.20	65.85	9.70	16.15	22.98
9	62.95	12.30	18.80	63.45	11.80	18.40	27.17

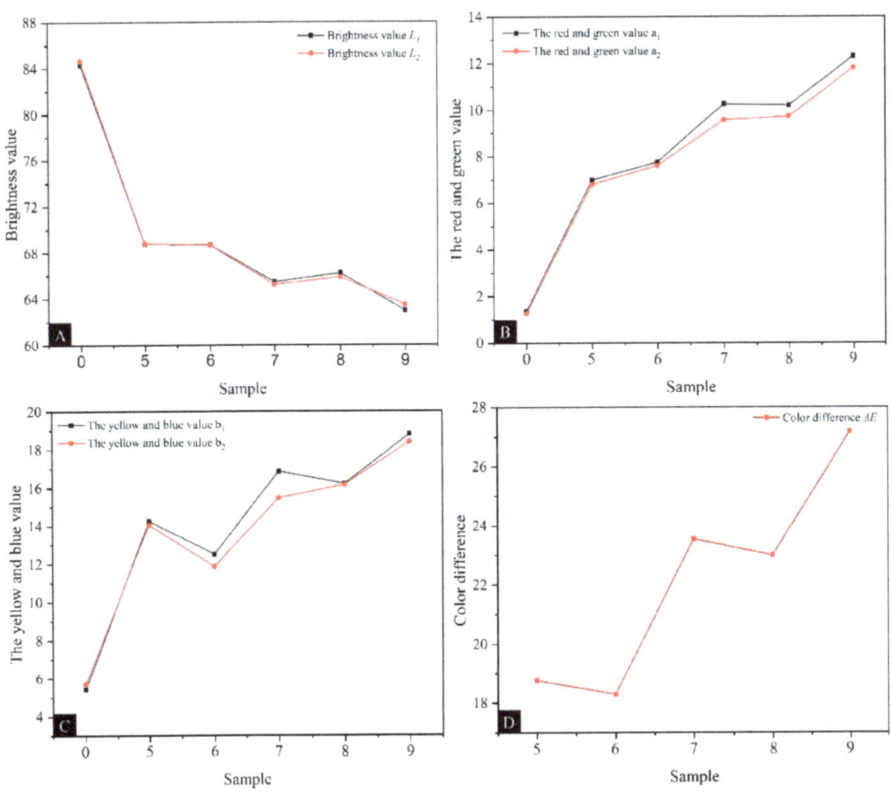

Figure 7. The effects of microcapsules with different M_T:M_C on the chromaticity value and color difference of the coating: (**A**) L value, (**B**) a value, (**C**) b value, and (**D**) ΔE.

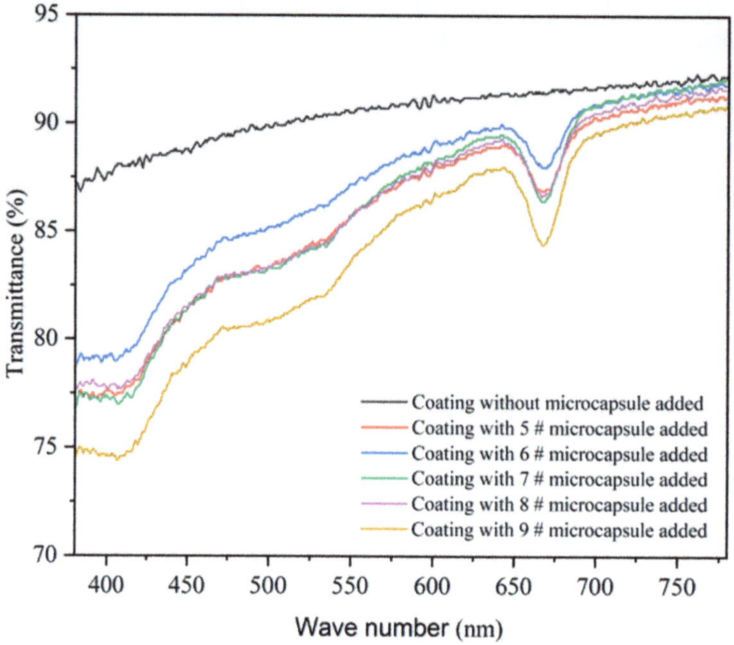

Figure 8. Effect of the M_T:M_C of microcapsules on the transmittance of the coating.

Table 11. Transmittance of coatings containing microcapsules with different M_T:M_C in the visible light band.

Sample (#)	Visible Light Transmittance (%)
0	90.44
5	85.81
6	87.05
7	86.00
8	85.96
9	84.08

3.4. Influence of Different M_T:M_C of Microcapsules on the Cold Liquid Resistance of the Coatings

Table 12 shows the cold liquid resistance of coatings prepared from microcapsules with different M_T:M_C. The cold resistance levels of the coating without microcapsules to citric acid, ethanol, and cleaning agents were level 2, level 3, and level 4, respectively. After adding microcapsule samples to the coating, the citric acid resistance level of the coating decreased to level 3. Compared to the tested area, the color differences in the test area were seen from multiple directions. This is because chitosan can dissolve in weak acid solutions [44], and as the shell material of microcapsules, chitosan is affected by citric acid, thereby reducing the cold liquid resistance of the coating to citric acid. Compared with the coating without microcapsules added, the ethanol cold liquid resistance of the coating was improved. When adding microcapsules of samples 5# and 6#, the ethanol resistance level of the coating was level 2. When adding microcapsules of samples 7#, 8#, and 9#, the ethanol resistance level of the coating was level 1. The cleaning agent cold liquid resistance of the coating showed an upward trend with the increase in M_T:M_C. The cleaning agent resistance level of the added samples 5# and 6# microcapsule coating was raised to level 3, and the cleaning agent resistance level of the added samples 7#, 8#, and 9# microcapsule coating was raised to level 2. This is because the added microcapsules have a protective effect on the coating, thereby improving its performance. When the M_T:M_C of microcapsules added to the coating was 3.0:1, 3.5:1, and 4.0:1, the antibacterial coating prepared had excellent cold liquid resistance performance.

Table 12. Cold-resistant liquid grade of coatings containing microcapsules with different M_T:M_C.

Sample (#)	Cold Liquid Resistance Level (Level)		
	Citric Acid	Ethanol	Cleaning Agents
0	2	3	4
5	3	2	3
6	3	2	3
7	3	1	2
8	3	1	2
9	3	1	2

3.5. The Influence of Different M_T:M_C of Microcapsules on the Mechanical Properties of the Coatings

Table 13 shows the fracture elongation of the coating prepared by microcapsules with different M_T:M_C. The fracture elongation of the coating without microcapsules added was 24.77%. After adding microcapsules, the fracture elongation of the coating showed a trend of first increasing and then decreasing and tended to stabilize. When adding the 7# microcapsule sample, the fracture elongation of the microcapsule coating reached its maximum value of 13.26%, at which point the microcapsule M_T:M_C was 3.0:1. The minimum fracture elongation of the coating was 7.24%, corresponding to the microcapsule M_T:M_C of 2.0:1. This is because the addition of microcapsules enhances the brittleness of the coating, thereby reducing its tensile resistance [45,46].

Table 13. Effect of microcapsules with different M_T:M_C on the mechanical properties of the coatings.

Sample (#)	Fracture Elongation (%)	Roughness (μm)
0	24.77	0.227
5	7.24	2.426
6	10.97	2.764
7	13.26	1.533
8	8.40	1.657
9	8.39	1.796

From the changes in the roughness values of the coating shown in Table 13, compared with the coating without microcapsules, the addition of microcapsules increased the roughness of the coating. As the microcapsule M_T:M_C increased, the roughness showed a fluctuating trend of first increasing, then decreasing, and then increasing again. When sample 7# was added to the coating, the roughness reached a minimum value of 1.533 μm, corresponding to the microcapsule M_T:M_C of 3.0:1. When adding sample microcapsules 5# and 6#, the roughness was 2.426 μm and 2.764 μm, and the microcapsule M_T:M_C were 2.0:1 and 2.5:1, respectively. This is because the aggregation phenomenon of microcapsules in samples 5# and 6# is more severe, which can easily reduce the dispersibility of microcapsules in waterborne coatings, causing more protrusions in the coating.

3.6. Effect of Microcapsules with Different M_T:M_C on the Antibacterial Performance of the Coatings

As shown in Table 14, compared to the colony recovery rate of the pure coating, the colony recovery rate of the coating with microcapsules added was significantly reduced. Figure 9 shows the trend of the antibacterial activity of microcapsules with different M_T:M_C against two types of bacteria in the coating. For *Escherichia coli*, as the microcapsule M_T:M_C increased, the antibacterial rate showed a trend of increasing, decreasing, and then increasing. When the added microcapsule M_T:M_C was 3.0:1, the antibacterial rate reached the maximum value of 67.14%. When the added microcapsule M_T:M_C was 2.5:1, the minimum antibacterial rate of the coating was 58.36%. This is because the coverage rate of microcapsules with M_T:M_C of 2.5:1 is lower, and the content of antibacterial agents in the core material is lower, so the antibacterial rate of the coating is lower. The antibacterial rate of the prepared coating against *Staphylococcus aureus* first increased and then decreased. When adding 7# microcapsules, the maximum antibacterial rate against *Staphylococcus aureus* was 68.39%, and the corresponding microcapsule M_T:M_C was 3.0:1. When the microcapsule M_T:M_C added to the coating was 2.0:1, the antibacterial rate reached the minimum value of 59.95%. This is because the core material content of microcapsules with M_T:M_C of 2.0:1 is relatively low, and the encapsulation rate of microcapsules is low. Therefore, there are fewer sustained-release antibacterial agents for the core material, and the antibacterial rate of the coating is lower. The antibacterial rate of the coating against *Staphylococcus aureus* was generally higher than that against *Escherichia coli*. This is because the effective antibacterial component coumarin compounds in the core material *Toddalia asiatica* (L.) Lam extract have better antibacterial effects on *Staphylococcus aureus* [47]. When the added microcapsule M_T:M_C was 3.0:1 (sample 7#), the antibacterial coating prepared had the highest antibacterial rates against *Escherichia coli* and *Staphylococcus aureus*.

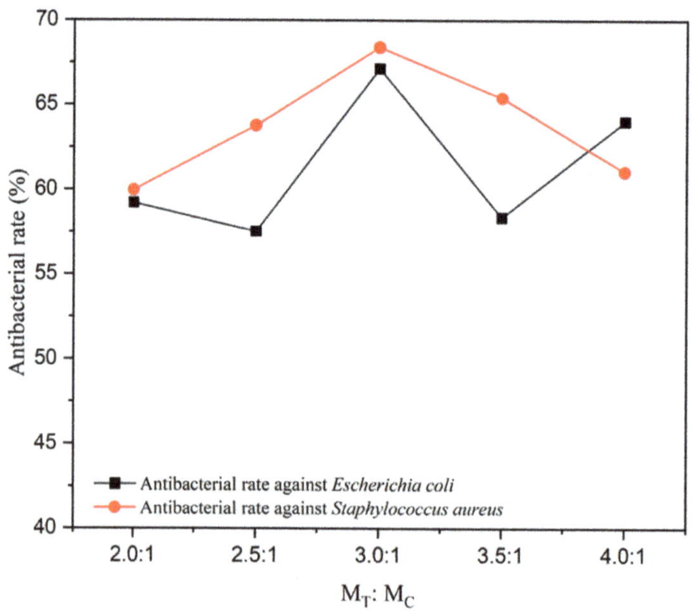

Figure 9. Antibacterial rate of coatings containing microcapsules with different M_T:M_C.

Table 14. The average number of recovered colonies and antibacterial rate of the coating-added microcapsules with different M_T:M_C.

Sample (#)	Average Number of Recovered *Escherichia Coli* (CFU·Piece^{-1})	Antibacterial Rate against *Escherichia Coli* (%)	Average Number of Recovered *Staphylococcus Aureus* (CFU·Piece^{-1})	Antibacterial Rate against *Staphylococcus Aureus* (%)
0	353	-	367	-
5	144	59.21	140	59.95
6	150	57.51	133	63.76

Table 14. *Cont.*

Sample (#)	Average Number of Recovered *Escherichia Coli* (CFU·Piece^{-1})	Antibacterial Rate against *Escherichia Coli* (%)	Average Number of Recovered *Staphylococcus Aureus* (CFU·Piece^{-1})	Antibacterial Rate against *Staphylococcus Aureus* (%)
7	116	67.14	116	68.39
8	147	58.36	127	65.40
9	127	64.02	143	61.04

4. Conclusions

The biggest influencing factor for preparing microcapsules was M_T:M_C. The optimal preparation process parameters for microcapsules were M_T:M_C of 3.0:1, a microencapsulation pH value of 7, and a reaction temperature of 50 °C. When the addition content was 5.0%, as the M_T:M_C of the added microcapsules increased, the glossiness showed a trend of first increasing and then decreasing. As the microcapsule M_T:M_C increased, the color difference value generally increased. The overall visible light transmittance increased and then decreased. The level of citric acid resistance of the coating decreased, while the level of ethanol and cleaning agent resistance increased, with the most significant improvement in cleaning agent resistance. The fracture elongation increased and then decreased, and the roughness value gradually increased. As the microcapsule M_T:M_C increased, the antibacterial rate against *Escherichia coli* increased, decreased, and then increased again. The antibacterial rate against *Staphylococcus aureus* increased and then decreased. When the microcapsule M_T:M_C was 3.0:1, the antibacterial rates of the coating against *Escherichia coli* and *Staphylococcus aureus* reached their maximum values, which were 67.14% and 68.39%, respectively. Based on the properties of the coating, the waterborne coating prepared with a microcapsule M_T:M_C of 3.0:1 exhibited excellent performance. The glossiness was 7.33 GU, the loss rate was 60.44%, the color difference ΔE was 23.53, the visible light transmittance was 86.00%, and the resistance to citric acid, ethanol, and cleaning agents was grade 3, grade 1, and grade 1, respectively. The fracture elongation was 13.26%, and the roughness was 1.533 μm. The antibacterial rates against *Escherichia coli* and *Staphylococcus aureus* were 67.14% and 68.39%, respectively. The preparation of *Toddalia asiatica* (L.) Lam extracts into microcapsules can overcome the problems of the poor heat resistance, poor stability, and short shelf life of general natural antibacterial agents, while also endowing antibacterial agents with sustained-release properties. Compared to commonly used shell materials such as melamine resin and urea formaldehyde resin, chitosan is more environmentally friendly, the drugs used are non-toxic, the recycling of production waste is easier, and it is more environmentally friendly. Chitosan and *Toddalia asiatica* (L.) Lam extracts both have antibacterial properties, and their synergistic effect can effectively enhance the antibacterial performance of the coating. The results effectively improved the antibacterial rate of waterborne coatings, providing a new approach for the application of d *Toddalia asiatica* (L.) Lam extract in waterborne coatings.

Author Contributions: Conceptualization and methodology and writing—review and editing, Y.Z.; validation, resources, and data management, Y.W.; formal analysis, investigation, and supervision, X.Y. All authors have read and agreed to the published version of the manuscript.

Funding: This project was partly supported by the Natural Science Foundation of Jiangsu Province (BK20201386).

Institutional Review Board Statement: Not applicable.

Informed Consent Statement: Not applicable.

Data Availability Statement: Data are contained within the article.

Conflicts of Interest: The authors declare that there are no conflicts of interest.

References

1. Zhang, E.L.; Zhao, X.T.; Hu, J.L.; Wang, R.X.; Fu, S.; Qin, G.W. Antibacterial metals and alloys for potential biomedical implan *Bioact. Mater.* **2021**, *6*, 2569–2612. [CrossRef] [PubMed]
2. Hu, W.; Fu, W.J.; Zhao, Y. Optimal design of the traditional Chinese wood furniture joint based on experimental and numeric method. *Wood Res.* **2024**, *69*, 50–59. [CrossRef]
3. Zhang, X.M.; Qin, M.; Xu, M.J.; Miao, F.Y.; Merzougui, C.; Zhang, X.Y.; Wei, Y.; Chen, W.Y.; Huang, D. The fabrication antibacterial hydrogels for wound healing. *Eur. Polym. J.* **2021**, *146*, 110268. [CrossRef]
4. Liu, Y.; Hu, W.; Kasal, A.; Erdil, Y.Z. The state of the art of biomechanics applied in ergonomic furniture design. *Appl. Sci.* **202** *13*, 12120. [CrossRef]
5. Jia, B.; Li, G.W.; Cao, E.R.; Luo, J.L.; Zhao, X.; Huang, H.Y. Recent progress of antibacterial hydrogels in wound dressings. *Mat Today Biol.* **2023**, *19*, 100582. [CrossRef] [PubMed]
6. Egghe, T.; Morent, R.; Hoogenboom, R.; De Geyter, N. Substrate-independent and widely applicable deposition of antibacteri coatings. *Trends Biotechnol.* **2023**, *41*, 63–76. [CrossRef] [PubMed]
7. Zhang, N.; Xu, W.; Tan, Y. Multi-attribute hierarchical clustering for product family division of customized wooden doo *BioResources* **2023**, *18*, 7889–7904. [CrossRef]
8. Sang, R.J.; Yang, S.Q.; Fan, Z.X. Effects of MDF substrate surface coating process on UV inkjet print quality. *Coatings* **2023**, *13*, 97 [CrossRef]
9. Shao, L.Y.; Xi, Y.W.; Weng, Y.X. Recent Advances in PLA-Based Antibacterial Food Packaging and Its Applications. *Molecul* **2022**, *27*, 5953. [CrossRef]
10. Chen, M.X.; Hu, Q.; Wang, X.Y.; Zhang, W. A review on recent trends of the antibacterial nonwovens air filter material Classification, fabrication, and application. *Sep. Purif. Technol.* **2024**, *330*, 125404. [CrossRef]
11. Liang, X.K.; Dai, R.; Chang, S.C.; Wei, Y.Q.; Zhang, B. Antibacterial mechanism of biogenic calcium oxide and antibacterial activi of calcium oxide/polypropylene composites. *Colloids Surf. A Physicochem. Eng. Asp.* **2022**, *650*, 129446. [CrossRef]
12. Li, X.X.; Gao, H.S.; Zhou, L.; Nie, Y. Screening Ionic Liquids by the COSMO-RS Method for the Preparation of Antibacteri Cellulose Fibers. *ACS Sustain. Chem. Eng.* **2021**, *9*, 15525–15536. [CrossRef]
13. Qi, Y.Q.; Liu, G.Y.; Zhang, Z.; Zhou, Z.W. Optimization of green extraction process of Cinnamomum camphora fruit dye and i performance by response surface methodology. *BioResources* **2023**, *18*, 4916–4934. [CrossRef]
14. Qi, Y.Q.; Zhang, Z.Q.; Sun, Y.; Shen, L.M.; Han, J.L. Study on the process optimization of peanut coat pigment staining of popla wood. *Forests* **2024**, *15*, 504. [CrossRef]
15. Sang, R.J.; Yang, F.; Fan, Z.X. The effect of water-based primer pretreatment on the performance of water-based inkjet coatings o wood surfaces. *Coatings* **2023**, *13*, 1649. [CrossRef]
16. Qi, Y.Q.; Zhou, Z.W.; Xu, R.; Dong, Y.T.; Zhang, Z.Q.; Liu, M.J. Effect of NaOH pretreatment on permeability and surfac properties of three wood species. *ACS Omega* **2023**, *8*, 40362–40374. [CrossRef] [PubMed]
17. Yang, S.Y.; Zhang, F.; Zhao, J.P.; Kang, H.J.; Ke, Q. Synthesis of inorganic/organic hybrid-shell antibacterial polyurea microcapsule loaded with Ag/TiO_2 nanoparticles. *Colloids Surf. A Physicochem. Eng. Asp.* **2024**, *691*, 133814. [CrossRef]
18. Hu, W.G.; Wan, H. Comparative study on weathering durability properties of phenol formaldehyde resin modified sweetgun and southern pine specimens. *Maderas. Cienc. Tecnol.* **2022**, *24*, 17. [CrossRef]
19. Liu, Q.Q.; Gu, Y.T.; Xu, W.; Lu, T.; Li, W.N.; Fan, H.B. Compressive properties of green velvet material used in mattress beddin *Appl. Sci.* **2021**, *11*, 11159. [CrossRef]
20. Gu, Y.T.; Zhang, J.L. Fatigue performance of natural and synthetic rattan strips subjected to cyclic tensile loading. *Forests* **2022**, *1* 76. [CrossRef]
21. Gu, Y.T.; Zhang, J.L. Tensile Properties of Natural and Synthetic Rattan Strips Used as Furniture Woven Materials. *Forests* **2020**, *1* 1299. [CrossRef]
22. Li, P.; Lei, K.; Ji, L.S. Characterization of the complete chloroplast genome of *Toddalia asiatica* (L.) Lam. *Mitochondrial DNA B* **202** *6*, 1650–1651. [CrossRef] [PubMed]
23. Zeng, Z.; Tian, P.; Feng, J.; Yang, N.A.; Yuan, L. A systematic review on traditional medicine *Toddalia asiatica* (L.) Lam.: Chemistr and medicinal potential. *Saudi Pharm. J.* **2021**, *29*, 781–798. [CrossRef] [PubMed]
24. He, N.; Wang, P.Q.; Wang, P.Y.; Ma, C.Y.; Kang, W.Y. Antibacterial mechanism of chelerythrine isolated from root of *Toddali asiatica* (Linn) Lam. *BMC Complement. Altern. Med.* **2018**, *18*, 261. [CrossRef] [PubMed]
25. Duraipandiyan, V.; Ignacimuthu, S. Antibacterial and antifungal activity of Flindersine isolated from the traditional medicina plant, *Toddalia asiatica* (L.) Lam. *J. Ethnopharmacol.* **2009**, *123*, 494–498. [CrossRef] [PubMed]
26. Liu, Q.Q.; Gao, D.; Xu, W. Effect of paint process on the performance of modified poplar wood antique. *Coating* **2021**, *11*, 1174 [CrossRef]
27. Xu, W.; Fang, X.Y.; Han, J.T.; Wu, Z.H.; Zhang, J.L. Effect of coating thickness on sound absorption property of four wood species commonly used for piano soundboards. *Wood Fiber Sci.* **2020**, *52*, 28–43. [CrossRef]
28. Qiu, H.Y.; Xiao, X.H.; Li, G.K. Separation and purification of furanocoumarins from *Toddalia asiatica* (L.) Lam. using microwave assisted extraction coupled with high-speed counter-current chromatography. *J. Sep. Sci.* **2012**, *35*, 901–906. [CrossRef]
29. Meng, Q.Y.; Zhong, S.L.; Wang, J.; Gao, Y.; Cui, X.J. Advances in chitosan-based microcapsules and their applications. *Carbohydr Polym.* **2023**, *300*, 120265. [CrossRef]

GB/T 11186.3-1989; Methods for Measuring the Color of Coatings—Part 3: Calculation of Color Difference. Standardization Administration of the People's Republic of China: Beijing, China, 1989.

GB/T 4893.6-2013; Testing of Physical and Chemical Properties of Furniture Surface Paint Films—Part 6: Gloss Determination Method. Standardization Administration of the People's Republic of China: Beijing, China, 2013.

GB/T 4893.1-2021; Test of Surface Coatings of Furniture—Part 1: Determination of Surface Resistance to Cold Liquids. Standardization Administration of the People's Republic of China: Beijing, China, 2021.

GB/T 21866-2008; Test Method and Effect for Antibacterial Capability of Paints Film. Standardization Administration of the People's Republic of China: Beijing, China, 2008.

GB/T 4789.2-2022; Microbiological Examination of Food Hygiene—Aerobic Plate Count. Standardization Administration of the People's Republic of China: Beijing, China, 2022.

Han, Y.; Yan, X.X.; Tao, Y. Effect of Number of Impregnations of *Microberlinla* sp. with Microcapsule Emulsion on the Performance of Self-Repairing Coatings on Wood Surfaces. *Coatings* **2022**, *12*, 989. [CrossRef]

Ciftci, N.; Sargin, I.; Arslan, G.; Arslan, U.; Okudan, A. Ascorbic Acid Adsorption-Release Performance and Antibacterial Activity of Chitosanter (GMA-MA-NTBA) Polymer Microcapsules. *J. Polym. Environ.* **2020**, *28*, 2277–2288. [CrossRef]

Chen, Z.; Ni, Y.Z.; Zhang, W.W.; Wang, Z.Y.; Ma, R.J.; Jia, Z.H.; Chen, K.L. Fabrication of multiple protective fabric based on ZnO/chitosan composite microcapsules via thiol-ene click modification. *Cellulose* **2023**, *30*, 8023–8036. [CrossRef]

Huang, N.; Yan, X.X.; Han, Y. Preparation of Melamine-Formaldehyde Resin/Rice Husk Powder Coated Epoxy Resin Microcapsules and Effects of Different Microcapsule Contents on the Properties of Waterborne Coatings on Tilia europaea Surface. *Coatings* **2022**, *12*, 1213. [CrossRef]

Sacco, P.; Pedroso-Santana, S.; Kumar, Y.; Joly, N.; Martin, P.; Bocchetta, P. Ionotropic Gelation of Chitosan Flat Structures and Potential Applications. *Molecules* **2021**, *26*, 660. [CrossRef] [PubMed]

Zhou, J.C.; Xu, W. A fast method to prepare highly isotropic and optically adjustable transparent wood-based composites based on interface optimization. *Ind. Crops Prod.* **2024**, *218*, 118898. [CrossRef]

Wang, C.; Zhou, Z.Y. Optical properties and lampshade design applications of PLA 3D printing materials. *BioResources* **2023**, *18*, 1545–1553. [CrossRef]

Liu, Q.Q.; Gao, D.; Xu, W. Effect of polyurethane non-transparent coating process on paint film performance applied on modified poplar. *Coating* **2022**, *12*, 39. [CrossRef]

Zhou, J.C.; Xu, W. Toward interface optimization of transparent wood with wood color and texture by silane coupling agent. *J. Mater.* **2022**, *57*, 5825–5838. [CrossRef]

Wang, J.L.; Zhuang, S.T. Chitosan-based materials: Preparation, modification and application. *J. Clean. Prod.* **2022**, *335*, 131825. [CrossRef]

Hu, W.G.; Chen, B.R.; Zhang, T.X. Experimental and numerical studies on mechanical behaviors of beech wood under compressive and tensile states. *Wood Res.* **2021**, *66*, 27–37. [CrossRef]

Hu, W.G.; Liu, N. Numerical and Optimal Study on Bending Moment Capacity and Stiffness of Mortise-and-Tenon Joint for Wood Products. *Forests* **2020**, *11*, 501. [CrossRef]

Zeki, N.M.; Mustafa, Y.F. Natural linear coumarin-heterocyclic conjugates: A review of their roles in phytotherapy. *Fitoterapia* **2024**, *175*, 105929. [CrossRef] [PubMed] *coatings*

MDPI

Article

Antimicrobial Efficacy of Sodium Dichloroisocyanurate Washing and Coating for Reduction of Foodborne Pathogens on Fresh Blackberries

Syed Ammar Hussain [1], Majher I. Sarker [2,*] and Tony Z. Jin [1,*]

[1] Microbial and Chemical Food Safety Research Unit, Eastern Regional Research Center, Agricultural Research Service, U.S. Department of Agriculture, 600 E. Mermaid Lane, Wyndmoor, PA 19038, USA; syed.a.hussain@usda.gov

[2] Sustainable Biofuels and Co-Products Research Unit, Eastern Regional Research Center, Agricultural Research Service, U.S. Department of Agriculture, 600 E. Mermaid Lane, Wyndmoor, PA 19038, USA

* Correspondence: majher.sarker@usda.gov (M.I.S.); tony.jin@usda.gov (T.Z.J.)

Abstract

Microbial contamination of fresh berries poses a significant public health risk due to their delicate and porous structure. This study evaluated the antimicrobial efficacy of washing and coating with sodium dichloroisocyanurate (SDC) against *Escherichia coli*, *Listeria monocytogenes*, and *Salmonella enterica* on blackberries. Two treatment approaches were tested: washing in SDC solutions (500 ppm and 1000 ppm for 1, 4, and 8 min), and hemicellulose B-based coatings supplemented with SDC (1 and 4 min exposure). Our results demonstrated that microbial reduction was significantly influenced by SDC concentration and exposure duration; longer treatment times and higher SDC concentrations achieved greater reductions. Washing with 1000 ppm SDC for 8 min obtained reductions of 5.0 ± 0.22 Log CFU/mL for *E. coli* ($p < 0.001$), 3.6 ± 0.45 Log CFU/mL for *L. monocytogenes* ($p < 0.01$), and 4.5 ± 0.30 Log CFU/mL for *S. enterica* ($p < 0.001$). Coating treatments at 1000 ppm for 4 min achieved even greater reductions for *S. enterica*, reaching 6.8 ± 0.35 Log CFU/mL, which was statistically superior to the corresponding washing treatment ($p < 0.001$). These findings support SDC as a promising antimicrobial agent for blackberry decontamination using a washing or coating method, with potential applications in fresh produce safety.

Keywords: blackberries; sodium dichloroisocyanurate (SDC); washing; edible coatings; hemicellulose B; microbial reduction; foodborne pathogens

1. Introduction

The blackberry (*Rubus fruticosus*) constitutes one of the many widely distributed berries among the Rubus genus, which belongs to the Rosaceae family. The berries are made up of an agglomeration of drupes, each about 1–3 cm in diameter, which vary in color from green to crimson to black as they ripen. Blackberries have been savored for centuries, cherished for their delightful flavor [1]. Berries, notably blackberries, have recently garnered popularity as "superfoods" due to their high concentration of bioactive chemicals. These include vitamins A and C, sterols, carotenoids, terpenoids, and phenolic constituents, all of which provide significant health advantages while remaining low in calories [2,3]. Blackberries contain high concentrations of bioactive components such as anthocyanins, flavonols, ellagitannins, and phenolic acids [4–6]. These phenolic compounds are well known for their anti-inflammatory, antioxidant, antidiabetic, and antibacterial properties [7]. Furthermore, they are thought to play a role in the prevention of chronic oxidative stress, cancer, and cardiac illnesses such myocardial cardiac disease and a fatal stroke [8]. In addition to their nutritional and therapeutic value, blackberries are particularly vulnerable to microbial contamination due to their delicate uneven surface structure, high moisture content, and minimal postharvest processing. Their frequent consumption in raw form further elevates the risk of foodborne illness, as pathogens are not eliminated through cooking. Notably, blackberries have been implicated in past contamination events, including *Salmonella* outbreaks traced to fresh berries in the United States and Europe, reinforcing their relevance in food safety research [9,10].

Foodborne pathogens contribute significantly to global morbidity and mortality. Each year in the United States, 31 major foodborne pathogens cause roughly 9.4 million illnesses, 56,000 hospital stays, and 1300 fatalities [9]. The most common bacterial pathogens linked to foodborne diseases are *L. monocytogenes*, enterohemorrhagic *E. coli* and *S. enterica* [9,10]. Bacterial contamination may take place at any stage throughout the production chain, notably through contact with contaminated water and interactions with those handling food or gear throughout harvesting and post-harvest [11]. Safeguarding the microbiological integrity of blackberries is critical for lowering the risk of foodborne diseases and preserving consumer trust. Chemical interventions, radiation exposure, and biological measures have all been investigated as ways to decontaminate fresh food [12,13]. Several studies have demonstrated the effects of chemicals such as ozone [14], sodium hypochlorite [15], lactic acid [16], and chlorine dioxide [17], on reducing the bacterial populations in berries. However, these reductions have generally not been significant [15–18]. Consequently, there is a growing demand for antimicrobial treatments that are low-cost, safe for consumption and highly effective at controlling microbes including *L. monocytogenes*, *E. coli*, and *S. enterica*, all of which contribute significantly to foodborne illness.

Sodium dichloroisocyanurate (SDC), a member of the *N*-halamine group of compounds, is a recognized biocidal agent due to its broad-spectrum antibacterial properties. The *N*-halamine compounds are distinguished by the presence of nitrogen-halogen covalent bonds. The antibacterial activity of SDC is attributed to the +1 oxidation state of the halide atoms in its molecular structure [19]. SDC is the sodium salt of a chlorinated hydroxy triazine and is utilized as a source of free accessible chlorine in the form of hypochlorous acid. It is extensively used as a reliable source of chlorine for swimming pool disinfection and in the food industry. It is used to disinfect drinking water, particularly in crises, since it is a convenient source of free chlorine. More recently, it has been used as a source of chlorine for point-of-use water treatment in homes [20]. SDC is less corrosive than bleach, exhibits mild toxicity, and is cost-effective, safe for humans, and environmentally benign [21,22]. Prior research has demonstrated the efficacy of SDC against prevalent foodborne intestinal protozoa, including *Giardia lamblia*, *Entamoeba histolytica*, *Cyclospora*, *Cryptosporidium*, and *Microsporidia*, as well as food-borne bacteria such as *L. monocytogenes*, *E. coli* and *S. enterica* [19,23].

Hemicellulose B, a plant-derived biopolymer, has shown promise as edible coating material for food applications due to its excellent film-forming ability, and compatibility with antimicrobial agents [24,25]. In our previous research, we developed hemicellulose-based packaging films specifically for food safety enhancement [26,27]. These formulations demonstrated improved antimicrobial retention, with notable effectiveness in reducing food borne pathogen populations during short-duration treatments.

This study systematically evaluates both washing and coating strategies using SDC as an antimicrobial agent for microbial decontamination of blackberry surfaces. Specifically, it investigates the impact of SDC concentration and exposure duration on reducing populations of three foodborne pathogens (*L. monocytogenes, E. coli* and *S. enterica*).

2. Materials and Methods

2.1. Preparation of Bacterial Strains

Stock cultures of *L. monocytogenes* (strain 19111), *E. coli* (ATCC strain 20R2R, serotype O157:H7), and *S. enterica* (strain 53647, serotype Typhimurium) were sourced from the USDA-ARS ERRC collection in Wyndmoor, PA, USA. To preserve these bacterial strains, they were stored at −80 °C in tryptic soy broth (TSB; BBL, Sparks, MD, USA) containing 20% glycerol. To guarantee viability, each starting culture was grown for 24 h at 37 °C on tryptic soy agar (TSA; BBL) slants containing 0.6% yeast extract and stored at 4 °C for two to four weeks.

For experimental preparation, a loopful of each bacterial strain was transferred from TSA slants to individual tubes containing 20 mL TSB. The cultures were grown at 37 °C on the rotary shaker at 5.0× *g* for 18 h, reaching an estimated concentration of ~9 log CFU/mL. After incubation, the bacterial cultures underwent centrifugation at 10,000× *g* for 10 min, after which the supernatant was removed. The resulting pellets were washed twice with 2 mL of 0.1% peptone water and subsequently reconstituted in a volume of 20 mL of 0.1% peptone water for further experimental applications. All experiments were performed in triplicate (n = 3) for each bacterial strain and treatment condition to ensure reproducibility and statistical robustness.

2.2. Preparation of Antimicrobial Solution

SDC, sourced from Sigma-Aldrich (St. Louis, MO, USA), was dissolved in sterile deionized (DI) water in precise quantities to reach the required concentrations for experimental use. The solutions were formulated at room temperature and left undisturbed for 24 h before being utilized in experiments.

2.3. Preparation of Coating Solution

Hemicellulose B was extracted from corn bran in our laboratory following the procedure detailed in our previous publications [24,26]. The coating solution was prepared by dispersing 1% (w/v) hemicellulose B into a 500 mL glass container containing sterilized deionized (DI) water supplemented with SDC at concentrations of either 500 ppm or 1000 ppm. The solution was mixed thoroughly to achieve homogeneity. For control treatments, sterilized DI water without SDC was used following the same protocol.

2.4. Disk Diffusion Assay

The disk diffusion assay was performed by inoculating TSA plates with ~8 log CFU of individual strains of *L. monocytogenes, E. coli*, and *S. enterica* (Table 1). After inoculation, the plates were left to air dry for 5 min. A 6 mm paper disk was carefully placed on each plate, and 100 µL of SDC at concentrations of either 250 ppm or 1000 ppm was applied to the disks. The plates were then incubated for 24 to 48 h, after which the zones of inhibition surrounding the disks were measured in millimeters (mm).

Table 1. Zones of inhibition for the different bacterial strains *E. coli, L. monocytogenes*, and *S. enterica* on TSA plates after treatment with SDC (1000 ppm). Each sample underwent testing in duplicate.

	Zone of Inhibition (Diameter in mm)		
Bacterial Pathogen	**Trial 1**	**Trial 2**	**Average**
E. coli	8	8.5	8.25
L. monocytogenes	9.5	9	9.25
S. enterica	8.5	8.25	8.37

2.5. Preparation of Blackberries and Inoculum

Fresh blackberries were purchased from neighborhood stores, refrigerated at 4 °C, and consumed within a week. Berries selected for analysis measured 1–2 cm in diameter and weighed around 7–8 g. A total of 80 blackberries underwent an initial rinse with water, followed by treatment with 70% ethyl alcohol spray to eliminate microorganisms. Afterward, the berries were placed inside a biosafety cabinet and allowed to dry for 2 h.

To inoculate the blackberries with microbial contaminants, they were submerged in 100 mL of an inoculum solution, which consisted of 2 mL of each foodborne pathogens thoroughly mixed with 98 mL of 0.1% peptone water (approximately 6–7 log CFU/mL) to ensure proper suspension and distribution. After a 1 min immersion, the berries were transferred back to the biosafety cabinet and left to dry at 22 $\pm$ 2 °C for an additional 2 h to facilitate microbial attachment.

2.6. Treatment of Blackberries and Microbiological Analysis

Two treatment strategies were applied to inoculated blackberry samples: washing in SDC solutions and application of a hemicellulose B-based coating.

For SDC solution treatments, twenty blackberries per group were immersed in 500 mL of sterilized deionized (DI) water containing SDC at concentrations of either 500 ppm or 1000 ppm. The samples were agitated at room temperature (22 °C) for exposure durations of 1, 4, or 8 min. Control groups were processed identically using sterile DI water without SDC.

Following treatment, ten blackberries from each group were transferred into individual Whirl-Pak® bags with 10 mL of 0.1% peptone water and homogenized for one minute using a stomacher. Serial dilutions were performed, and aliquots were plated onto selective media to identify three pathogens, PALCAM agar (BBL/Difco, Sparks, MD, USA) (with selective supplement) for *L. monocytogenes*, Xylose Lysine Tergitol-4 (XLT4) agar for *S. enterica*, Tryptic Soy Agar (TSA, BBL/Difco) for *E. coli*.

For coating treatments, 20 blackberries were treated with a coating solution composed of 1% (w/v) hemicellulose B dispersed in DI water containing either 500 ppm or 1000 ppm of SDC. Samples were agitated at room temperature for durations of 1 or 4 min only. Sterile DI water was used for controls.

Microbiological analysis for the coating treatment targeted *S. enterica* exclusively. Ten blackberries per group were homogenized as described above, followed by serial dilutions and plating on XLT4 agar. Plates were incubated at 37 °C for up to 48 h, after which the log colony forming units (CFU) were determined. All microbiological assays were conducted in triplicate to ensure experimental reproducibility.

2.7. Statistical Analysis

The results from triplicate tests were expressed as the mean $\pm$ standard deviation. Statistical analysis was performed using GraphPad Prism 7.0 (GraphPad Software, Inc., San Diego, CA, USA), employing a one-way analysis of variance (ANOVA), followed by a post hoc Tukey's test. Before applying ANOVA, the normality of the data was checked using the Shapiro–Wilk test. If the data did not meet the assumptions for normal distribution, appropriate adjustments or non-parametric methods were considered. Statistically significant differences between treatment means were indicated by asterisks (* $p < 0.05$; ** $p < 0.01$; *** $p < 0.001$).

3. Results and Discussions

To improve the microbiological safety of blackberries, which are highly prone to contamination due to their delicate surface, this study evaluated two post-inoculation treatments: SDC washing and coating. Washing treatments targeted *L. monocytogenes*, *E. coli*, and *S. enterica* over varied exposure times, while the coating approach focused solely on *S. enterica* at shorter durations. Pathogen reductions were assessed via selective media to compare treatment efficacy.

3.1. Disk Diffusion Assays and Selection of SDC Concentration

Disk diffusion assays were conducted to evaluate the antimicrobial effectiveness of SDC on different bacterial strains. When 250 ppm of SDC was applied to sample discs on TSA plates, no inhibition zones were observed across the tested strains. However, increasing the concentration to 1000 ppm yielded inhibition zones with average diameters reaching up to 9.25 mm for the tested strains, as shown in Table 1. Consequently, for subsequent experiments on blackberries application, the SDC concentration was adjusted to 500 ppm, under the assumption that 250 ppm would be insufficient. Furthermore, the concentration was raised to 500 ppm for experiments with blackberries to account for the fruit's porous surface, which could potentially limit the antimicrobial's effectiveness.

3.2. Treatment of Blackberry Surfaces with SDC Solutions

The effectiveness of SDC in lowering populations of *L. monocytogenes*, *E. coli*, and *S. enterica* was assessed at two doses (500 ppm and 1000 ppm) as well as across three exposure durations (1 min, 4 min, & 8 min). The results demonstrated that decline in bacterial population was impacted by both the treatment concentration and the length of exposure, with longer durations and higher concentrations leading to greater decreases in microbial load (Figures 1–3).

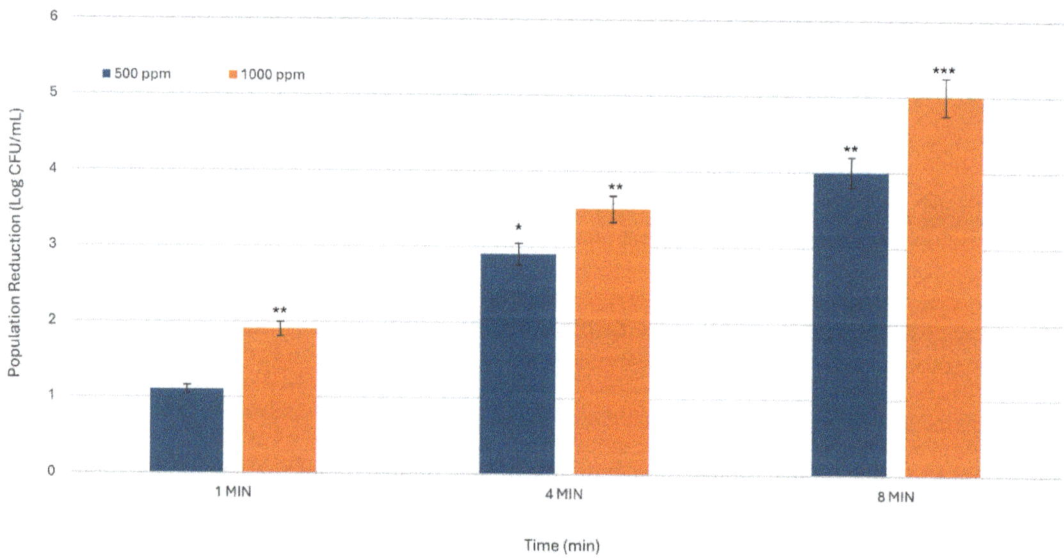

Figure 1. The inhibitory activity of SDC against *E. coli* is dependent on both time and concentration. The X-axis denotes treatment duration in minutes, while the Y-axis indicates the reduction in bacterial population, expressed in Log CFU/mL. The figure presents the average results from three experiments conducted in duplicate. Asterisks indicate statistically significant differences between the means of various treatments (* $p < 0.05$; ** $p < 0.01$; *** $p < 0.001$).

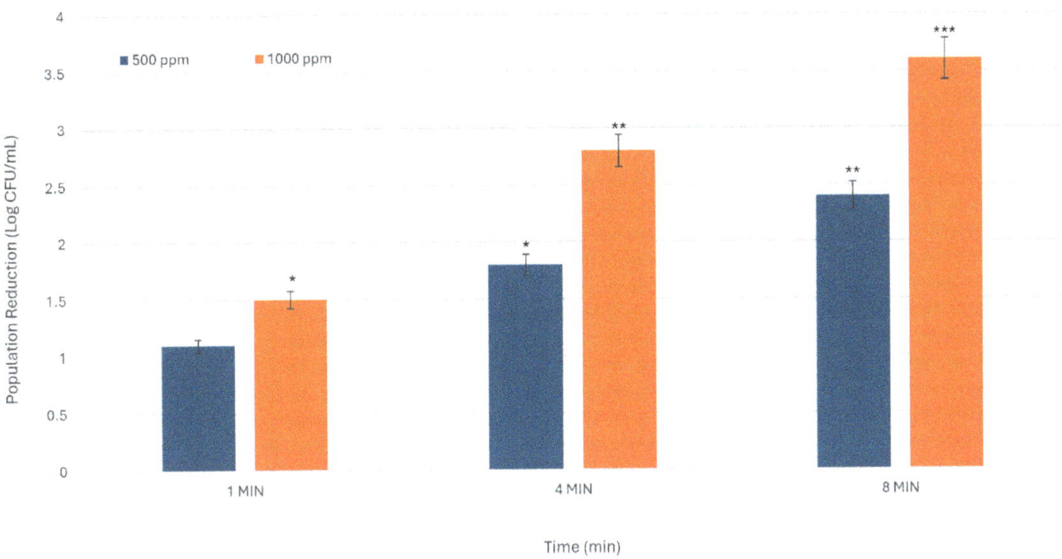

Figure 2. The inhibitory activity of SDC against *L. monocytogenes* is dependent on both time and concentration. The X-axis denotes treatment duration in minutes, while the Y-axis indicates the reduction in bacterial population, expressed in Log CFU/mL. The figure presents the average results from three experiments conducted in duplicate. Asterisks indicate statistically significant differences between the means of various treatments (* $p < 0.05$; ** $p < 0.01$; *** $p < 0.001$).

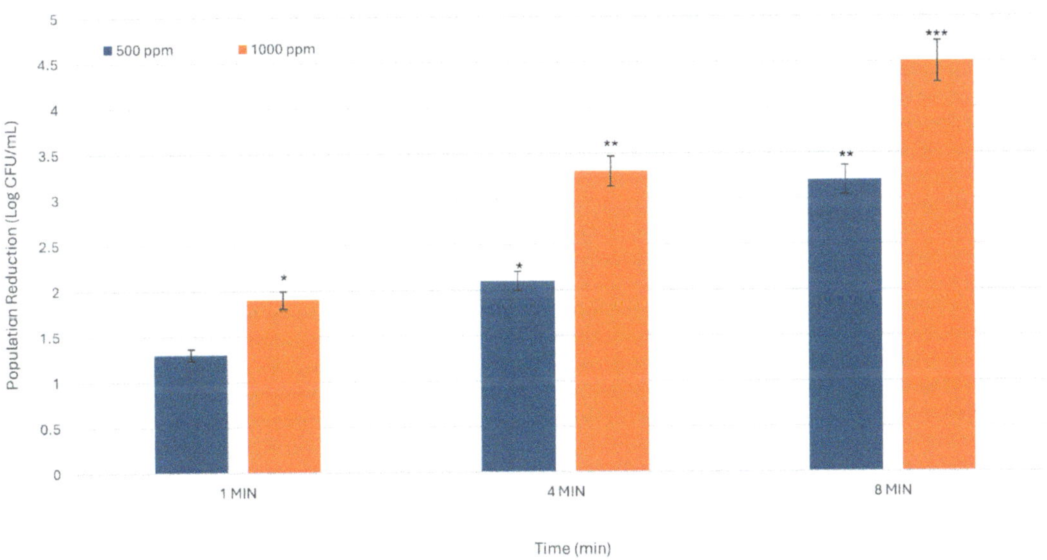

Figure 3. The inhibitory activity of SDC against *S. enterica* is dependent on both time and concentration. The X-axis denotes treatment duration in minutes, while the Y-axis indicates the reduction in bacterial population, expressed in Log CFU/mL. The figure presents the average results from three experiments conducted in duplicate. Asterisks indicate statistically significant differences between the means of various treatments (* $p < 0.05$; ** $p < 0.01$; *** $p < 0.001$).

For *E. coli*, microbial reduction at 500 ppm was recorded as 1.1, 2.9, and 4.0 Log CFU/mL at exposure times of 1, 4, and 8 min, respectively. When tested at a concentration of 1000 ppm, bacterial levels showed a notable decline, with reductions of 1.9, 3.5, and 5.0 Log CFU/mL ($p < 0.001$) (Figure 1) in comparison to control group (i.e., sterile deionized water). These results demonstrate that higher concentrations and longer exposure times contribute to a more pronounced reduction in *E. coli* populations on blackberry surfaces. However, upon analysis of the data across two concentrations and three duration intervals, it was observed that a 52.6% greater reduction in microbial loads on the surface of blackberries could be attained by increasing the exposure time from 1 to 4 min, necessitating a lower dosage of 500 ppm SDC rather than 1000 ppm (2.9 Log CFU/mL at 500 ppm in 4 min; 1.9 Log CFU/mL at 1000 ppm in 1 min). In a comparable manner, a further reduction of 14.2% could be achieved with a 500 ppm solution, necessitating an extension of the exposure time by 4 min (4 Log CFU/mL at 500 ppm in 8 min; 3.5 Log CFU/mL at 1000 ppm in 4 min).

For *L. monocytogenes*, microbial reduction at 500 ppm was observed at 1.1, 1.8, and 2.4 Log CFU/mL for exposure durations of 1, 4, and 8 min, respectively. At 1000 ppm concentration, microbial levels reduced through 1.5, 2.8, and 3.6 Log CFU/mL ($p < 0.01$) over the designated time intervals when compared to the control samples (Figure 2). These results highlight the effectiveness of higher concentrations in reducing bacterial presence. These findings illustrate that increasing concentration and extending the exposure duration led to more substantial microbial reduction. Analyzing the correlation between concentration and exposure time, it was noted that a 20% and 60% greater reduction in *L. monocytogenes* on blackberry surfaces could be attained with a 500 ppm solution (instead of 1000 ppm), necessitating an increase in exposure time from 1 to 4 min and from 1 to 8 min, respectively. (1.8 and 2.4 Log CFU/mL at 500 ppm in 4 and 8 min, respectively; 1.5 Log CFU/mL at 1000 ppm in 1 min).

For *S. enterica*, microbial reductions at 500 ppm were recorded as 1.3, 2.1, and 3.2 Log CFU/mL at exposure times of 1, 4, and 8 min, respectively. When tested at 1000 ppm, microbial reductions were 1.91, 3.3, & 4.5 Log CFU/mL ($p < 0.001$), showing a notable increase relative to the corresponding control samples (Figure 3). These results highlight the effectiveness of higher concentrations in reducing bacterial presence. This pattern highlights the direct correlation between higher concentrations and extended exposure durations in achieving greater microbial reduction.

3.3. Treatment of Blackberry Surfaces with Hemicellulose-Based Coating Approach

The effectiveness of the hemicellulose B-based coating formulation, supplemented with SDC, was assessed for its ability to reduce *S. enterica* populations on blackberry surfaces. Treatments were performed using two concentrations of SDC (500 ppm and 1000 ppm) and evaluated at exposure times of 1 and 4 min.

At 500 ppm, microbial reductions were 3.2 and 5.4 Log CFU/mL for the 1 min and 4 min exposures, respectively. When tested at 1000 ppm, reductions increased to 5.2 and 6.8 Log CFU/mL ($p < 0.001$), across the same time intervals, indicating a clear enhancement in antimicrobial activity with higher concentration and longer contact time (Figure 4). These results demonstrate a strong positive correlation between treatment intensity (concentration and duration) and microbial reduction. Notably, the coating approach yielded substantial decreases in *S. enterica*, outperforming washing treatments under comparable conditions (Table 2).

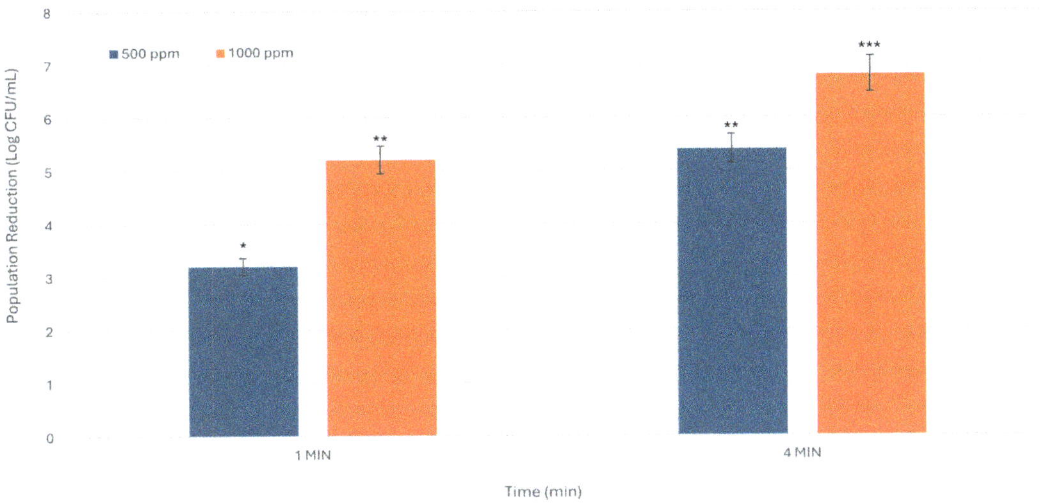

Figure 4. The inhibitory activity of SDC hemicellulose-based coating against *S. enterica* is dependent on both time and concentration. The X-axis denotes treatment duration in minutes, while the Y-axis indicates the reduction in bacterial population, expressed in Log CFU/mL. The figure presents the average results from three experiments conducted in duplicate. Asterisks indicate statistically significant differences between the means of various treatments (* $p < 0.05$; ** $p < 0.01$; *** $p < 0.001$).

Table 2. Demonstrates the best-performing SDC concentration and time for microbial reduction on blackberries. [a] Means are expressed as log CFU/mL ± SD. Asterisks to the left of the mean indicate a significant difference between treatment means. Significance was defined as * $p < 0.05$; ** $p < 0.01$; *** $p < 0.001$.

Bacterial Pathogen	Best-Performing SDC Concentration (ppm)	Best-Performing Exposure Time (min)	[a] Log Reduction (CFU/mL)
E. coli	1000 (Washing)	8	** 5.0 ± 0.21
L. monocytogenes	1000 (Washing)	8	* 3.6 ± 0.45
S. enterica	1000(Washing)	8	** 4.5 ± 0.11
S. enterica	1000 (Coating)	4	*** 6.8 ± 0.18

Overall, this intervention underscores the antimicrobial efficacy of SDC in reducing microbial contamination on blackberry surfaces through both washing and hemicellulose-based coating strategies. Across solution-based treatments, increased SDC concentration (from 500 ppm to 1000 ppm) and extended exposure times (up to 8 min) consistently improved microbial reductions for *L. monocytogenes*, *E. coli*, and *S. enterica*. These findings confirm that both concentration and contact time are key parameters for optimizing disinfection outcomes (Table 2), aligning with previous research highlighting the effectiveness of chlorine-based agents in fresh produce safety [28–30].

Interestingly, extended exposure of lower concentrations also yielded significant reductions ($p < 0.05$), suggesting operational flexibility depending on dosage availability and processing time constraints. For instance, the washing treatment with 500 ppm solution for 4 or 8 min could achieve comparable results, and in some cases superior reductions relative to shorter exposures at 1000 ppm. Such data offers practical insights for tailoring sanitization protocols across production environments, particularly where chemical minimization is desired without compromising efficacy [31,32].

Beyond solution washing, the hemicellulose B-based coating approach exhibited even greater efficacy, especially against *S. enterica*, with reductions reaching up to 6.8 Log CFU/mL at 1000 ppm within just 4 min ($p < 0.001$). This superior performance possibly stems from the polymer's capacity to retain SDC on the fruit surface, promoting prolonged contact and targeted antimicrobial release. On porous produce like blackberries, such delivery systems enhance overall pathogen control and may contribute to shelf-life extension while minimizing risk of pathogen rebound.

The notable performance of hemicellulose coatings builds on previous work demonstrating their potential in food packaging and surface decontamination applications [24–27]. As a biodegradable, edible film-forming material, hemicellulose B can serve as both an antimicrobial agent and as a sustainable alternative to synthetic polymers. Although hemicellulose does not possess any antimicrobial activity, its ability to integrate with active agents like SDC enhances retention of antimicrobial activity, protects fragile surfaces, and supports post-harvest handling protocols. By integrating hemicellulose-based coating treatments, this research advances a safer, more efficient framework for fresh produce sanitation and sets the stage for developing targeted edible antimicrobial systems for commercial applications.

Despite its strengths, this study has its limitations. Microbial analyses were performed immediately post-treatment without monitoring long-term pathogen suppression, shelf-life stability, or residual activity during storage. Coating treatments were evaluated only against *S. enterica*, leaving efficacy against *L. monocytogenes* and *E. coli* unaddressed in that format. Additionally, parameters such as drying time, uniform coating application, and cost-effectiveness were not investigated. The absence of sensory testing [33,34], also restricts conclusions regarding consumer experience and market viability. Given that SDC is a strong oxidizing agent, future studies should assess its potential to alter fruit appearance, taste, aroma, and nutritional properties, especially at higher concentrations. Furthermore, the physical and chemical properties of the hemicellulose B-based coating, such as viscosity, thickness, homogeneity, adherence to fruit surfaces, and stability under storage conditions were not characterized in this study. We acknowledge that these attributes are essential for interpreting antimicrobial performance and assessing practical feasibility, especially in regulatory and scalability contexts. This omission limits the ability to fully evaluate the coating's behavior in real-world applications. However, we note that this study was designed as a proof-of-concept, with a primary focus on demonstrating antimicrobial efficacy. The coating formulation was used to explore its potential as a delivery system for SDC. Although SDC is widely used and generally considered low in toxicity, its direct application to food surfaces warrants further safety evaluation to address potential risks such as skin irritation or respiratory effects at high exposure levels. The promising results observed here suggest that SDC-infused hemicellulose coatings could be integrated into commercial workflows, such as automated spray systems or edible film applications in modified atmosphere packaging. Such implementations may contribute to reducing microbial load during storage and distribution, extending shelf life, and offering a sustainable alternative to conventional antimicrobial interventions like chlorine rinses or ozone treatments. In addition, preliminary biodegradability testing confirmed that the hemicellulose-based coating is environmentally degradable, with full data to be presented in a forthcoming study. This supports its potential as a sustainable material for food safety applications.

To advance this work, future studies should explore broader pathogen coverage in coating systems, assess microbial stability under realistic storage conditions, and perform detailed sensory analyses across treated fruits. Comparative trials across different produce types, such as strawberries, grapes, or tomatoes would help verify transferability of findings. Optimization of formulation protocols, scale-up strategies, and regulatory compliance will be essential for industrial adoption. Specifically, it will be important to determine how long and at what concentration SDC remains on fruit surfaces after coating or washing, to assess the potential risk of ingesting harmful levels. Further research should also define optimal SDC concentrations and exposure durations to support practical application. As demonstrated in previous studies by Nazaruddin et al. [35], and Carpena et al. [36], integrating natural compounds into biodegradable matrices can enhance food safety through intelligent responsiveness and spoilage detection. Inspired by these approaches, combining SDC with natural antimicrobials or antioxidants may open new avenues for synergistic control and shelf-life enhancement in fresh produce management.

4. Conclusions

This study demonstrates the effectiveness of SDC in reducing microbial contamination on blackberries through two intervention strategies: washing and coatings. The antimicrobial activity of SDC was strongly influenced by concentration and contact time, with higher levels and longer exposures resulting in substantial reductions in L. *monocytogenes*, *E. coli*, and *S. enterica*. Notably, the hemicellulose-based coating approach achieved superior microbial reductions against *S. enterica* in shorter durations, suggesting enhanced retention and delivery of SDC on berry surfaces. This highlights the potential of coating-based systems for rapid and efficient pathogen control, particularly on delicate fruits like blackberries. SDC demonstrated promising antimicrobial efficacy under laboratory conditions, though its application in real-world produce safety requires further validation, including coating characterization, long-term stability, and comparative trials. The current study did not include sensory analysis, residue quantification, or simulation of commercial packaging and handling conditions. These limitations underscore the preliminary nature of the findings and highlight the need for future research to assess practical feasibility, safety, and scalability. Such efforts will be essential to determine whether SDC-infused coatings can serve as a sustainable and effective alternative to conventional antimicrobial interventions in fresh produce management.

Author Contributions: Conceptualization, S.A.H., M.I.S. and T.Z.J.; methodology, S.A.H. and M.I.S.; validation, S.A.H.; formal analysis, S.A.H.; investigation, S.A.H.; resources, M.I.S. and T.Z.J.; data curation, S.A.H., M.I.S. and T.Z.J.; writing—original draft, S.A.H.; writing—review and editing, S.A.H., M.I.S. and T.Z.J.; visualization, S.A.H.; supervision, M.I.S. and T.Z.J.; funding acquisition, T.Z.J. All authors have read and agreed to the published version of the manuscript.

Funding: This research was funded by the U.S. Department of Agriculture-Agriculture Research Services.

Institutional Review Board Statement: Not applicable.

Data Availability Statement: Data is contained within the article.

Acknowledgments: The authors acknowledge Anita Parameswaran (ERRC-ARS-USDA) for providing technical assistance.

Conflicts of Interest: The authors declare no conflicts of interest.

References

1. Verma, R.; Gangrade, T.; Punasiya, R.; Ghulaxe, C. *Rubus fruticosus* (Blackberry) use as an herbal medicine. *Pharmacogn. Rev.* **2014**, *8*, 101–104. [CrossRef]
2. D'Angelo, R.W.O.; Gonçalves, M.M.; Fachi, M.M.; Vilhena, R.d.O.; Pontarolo, R.; Maluf, D.F. UPLC–QToF-MS characterization of blackberry extracts of cultivars 'Tupy', 'Guarani', and 'Xavante': Development of extract-loaded niosomes. *Rev. Bras. Farmacogn.* **2020**, *30*, 519–527. [CrossRef]

3. Jara-Palacios, M.J.; Santisteban, A.; Gordillo, B.; Hernanz, D.; Heredia, F.J.; Escudero-Gilete, M.L. Comparative study of red ber pomaces (blueberry, red raspberry, red currant, and blackberry) as sources of antioxidants and pigments. *Eur. Food Res. Techn* **2019**, *245*, 1–9. [CrossRef]
4. Dai, J.; Patel, J.D.; Mumper, R.J. Characterization of blackberry extract and its antiproliferative and anti-inflammatory properti *J. Med. Food* **2007**, *10*, 258–265. [CrossRef]
5. Hager, T.J.; Howard, L.R.; Liyanage, R.; Lay, J.O.; Prior, R.L. Ellagitannin composition of blackberry as determined by HPLC-E MS and MALDI-TOF-MS. *J. Agric. Food Chem.* **2008**, *56*, 661–669. [CrossRef]
6. Oszmiański, J.; Nowicka, P.; Teleszko, M.; Wojdyło, A.; Cebulak, T.; Oklejewicz, K. Analysis of phenolic compounds ar antioxidant activity in wild blackberry fruits. *Int. J. Mol. Sci.* **2015**, *16*, 14540–14553. [CrossRef] [PubMed]
7. Folmer, F.; Basavaraju, U.; Jaspars, M.; Hold, G.; El-Omar, E.; Dicato, M.; Diederich, M. Anticancer effects of bioactive ber compounds. *Phytochem. Rev.* **2014**, *13*, 295–322. [CrossRef]
8. Zia-Ul-Haq, M.; Riaz, M.; De Feo, V.; Jaafar, H.Z.E.; Moga, M. *Rubus fruticosus* L.: Constituents, biological activities, ar health-related uses. *Molecules* **2014**, *19*, 10998–11029. [CrossRef]
9. Scallan, E.; Hoekstra, R.M.; Angulo, F.J.; Tauxe, R.V.; Widdowson, M.A.; Roy, S.L.; Jones, J.L.; Griffin, P.M. Foodborne illne acquired in the United States—Major pathogens. *Emerg. Infect. Dis.* **2011**, *17*, 7–15. [CrossRef] [PubMed]
10. Yang, W.; Chen, X.; Li, Y.; Guo, S.; Wang, Z.; Yu, X. Advances in pharmacological activities of terpenoids. *Nat. Prod. Commu* **2020**, *15*, 1934578X20903555. [CrossRef]
11. European Food Safety Authority (EFSA). Scientific opinion on the risk posed by pathogens in food of non-animal origin. *EFSA* **2013**, *11*, 3025. [CrossRef]
12. Zhang, W.; Zhao, W.; Sun, Y.; Sun, Y.; Wang, B.; Liu, M.; Qiu, Z.; Wang, Y.; Sun, Z.; Hu, P.; et al. Effects of multi-wavelengt ultraviolet radiation on the inactivation and reactivation of *E. coli* in a recirculating water system. *Aquac. Rep.* **2025**, *41*, 10268 [CrossRef]
13. Pérez-Lavalle, L.; Carrasco, E.; Valero, A. Strategies for microbial decontamination of fresh blueberries and derived produc *Foods* **2020**, *9*, 1558. [CrossRef]
14. Wu, V.C.H.; Kim, B. Effect of a simple chlorine dioxide method for controlling five foodborne pathogens, yeasts, and molds o blueberries. *Food Microbiol.* **2007**, *24*, 794–800. [CrossRef]
15. Goodburn, C.; Wallace, C.A. The microbiological efficacy of decontamination methodologies for fresh produce: A review. *Foo Control* **2013**, *32*, 418–427. [CrossRef]
16. Tadepalli, S.; Bridges, D.F.; Driver, R.; Wu, V.C.H. Effectiveness of different antimicrobial washes combined with freezing agains *E. coli* O157:H7, *Salmonella Typhimurium*, and *Listeria monocytogenes* inoculated on blueberries. *Food Microbiol.* **2018**, *74*, 34–3 [CrossRef]
17. Pangloli, P.; Hung, Y.-C. Reducing microbiological safety risk on blueberries through innovative washing technologies. *Foo Control* **2013**, *32*, 621–625. [CrossRef]
18. Nadeem, M.; Leaks, K.; Abdullah, A.; Adamson, F.J.; Shahid, M.A. Enhancing postharvest quality of blackberries: Impact c sonicated and microwave-assisted pasteurized edible coating gels at different storage temperatures. *Gels* **2025**, *11*, 243. [CrossRef
19. Long, W., III; Sarker, M.I.; Annous, B.A.; Paoli, G.C. Evaluation of sodium dichloroisocyanurate treatment on recovere concentrations of *Salmonella enterica*, *E. coli* O157:H7, and *Listeria monocytogenes* from cattle hide surfaces and culture mediur *J. Food Saf.* **2020**, *40*, e12834. [CrossRef]
20. WHO. Sodium Dichloroisocyanurate in Drinking Water. Available online: https://www.who.int/docs/default-source/wash documents/wash-chemicals/sodium-dichloroisocyanurate-2add-feb2008.pdf (accessed on 1 August 2025).
21. Dong, A.; Wang, Y.J.; Gao, Y.; Gao, T.; Gao, G. Chemical insights into antibacterial N-halamines. *Chem. Rev.* **2017**, *117*, 4806–486 [CrossRef]
22. Sarker, M.I.; Long, W.; Liu, C.K. Efficacy of aqueous solution of N-halamine to reduce microbiological contamination on cattl hides for meat safety with byproduct quality assurance. *J. Am. Leather Chem. Assoc.* **2020**, *115*, 330–336. [CrossRef]
23. Zou, Q.; Meng, W.; Wang, C.; Wang, T.; Liu, X.; Li, D. Sodium dichloroisocyanurate: Improving broiler health by reducin harmful microbial levels in the waterline. *Front. Vet. Sci.* **2023**, *10*, 1234949. [CrossRef]
24. da Silva Braga, R.; Poletto, M. Preparation and Characterization of Hemicellulose Films from Sugarcane Bagasse. *Materials* **202** *13*, 941. [CrossRef]
25. Zhao, Y.; Sun, H.; Yang, B.; Weng, Y. Hemicellulose-Based Film: Potential Green Films for Food Packaging. *Polymers* **202** *12*, 1775. [CrossRef]
26. Hussain, S.A.; Qi, P.X.; Sharma, B.K.; Yadav, M.P.; Mainali, K.; Jin, T.Z. Valorization of corn bran-derived carbohydrate polymer for developing biodegradable packaging films. *J. Polym. Environ.* **2025**, *33*, 2552–2566. [CrossRef]
27. Hussain, S.A.; Sharma, B.K.; Qi, P.X.; Yadav, M.P.; Jin, T.Z. Antimicrobial and Physicochemical Properties of Hemicellulose-Basec Films Incorporating Carvacrol. *Polymers* **2025**, *17*, 2073. [CrossRef]

Ma, J.W.; Huang, B.S.; Hsu, C.W.; Peng, C.W.; Cheng, M.L.; Kao, J.Y.; Way, T.D.; Yin, H.C.; Wang, S.S. Efficacy and safety evaluation of a chlorine dioxide solution. *Int. J. Environ. Res. Public Health* **2017**, *14*, 329. [CrossRef]
Malka, S.K.; Park, M.H. Fresh produce safety and quality: Chlorine dioxide's role. *Front. Plant Sci.* **2022**, *12*, 775629. [CrossRef]
Gerba, C.P. Quaternary ammonium biocides: Efficacy in application. *Appl. Environ. Microbiol.* **2015**, *81*, 464–469. [CrossRef]
Luu, P.; Chhetri, V.S.; Janes, M.E.; King, J.M.; Adhikari, A. Effectiveness of aqueous chlorine dioxide in minimizing food safety risk associated with *Salmonella*, *E. coli* O157:H7, and *Listeria monocytogenes* on sweet potatoes. *Foods* **2020**, *9*, 1259. [CrossRef]
Wang, F.; Xiang, L.; Leung, K.S.-Y.; Elsner, M.; Zhang, Y.; Guo, Y.; Pan, B.; Sun, H.; An, T.; Ying, G.; et al. Emerging contaminants: A One Health perspective. *Innovation* **2024**, *5*, 100612. [CrossRef]
Martín-Diana, A.B.; Rico, D.; Barry-Ryan, C.; Frías, J.M.; Mulcahy, J.; Henehan, G.T. Comparison of calcium lactate with chlorine as a washing treatment for fresh-cut lettuce and carrots: Quality and nutritional parameters. *JSFA Rep.* **2005**, *85*, 2260–2268. [CrossRef]
Nakano, H.; Pan, X. Effects of chlorine-based antimicrobial treatments on the microbiological qualities of selected leafy vegetables and wash water. *Food Sci. Technol. Res.* **2014**, *20*, 765–774. [CrossRef]
Nazaruddin, N.; Afifah, N.; Bahi, M.; Susilawati, S.; Sani, N.D.M.; Esmaeili, C.; Iqhrammullah, M.; Murniana, M.; Hasanah, U.; Safitri, E. A simple optical pH sensor based on pectin and Ruellia tuberosa L-derived anthocyanin for fish freshness monitoring. *F1000Research* **2021**, *10*, 422. [CrossRef]
Carpena, M.; Silva, A.; Barciela, P.; Perez-Vazquez, A.; Chamorro, F.; Cassani, L.; Barroso, M.F.; Xiao, J.; Prieto, M.A.; Simal-Gandara, J. Inclusion of Natural Anthocyanins as Food Spoilage Sensors. *Eng. Proc.* **2023**, *48*, 59. [CrossRef]

coatings

Article

Optimization of Preparation Process for Chitosan-Coated Pomelo Peel Flavonoid Microcapsules and Its Effect on Waterborne Paint Film Properties

Jinzhe Deng [1,2], Tingting Ding [1,2] and Xiaoxing Yan [1,2,*]

1 Co-Innovation Center of Efficient Processing and Utilization of Forest Resources, Nanjing Forestry University, Nanjing 210037, China; dengjinzhe@njfu.edu.cn (J.D.); dingtingting@njfu.edu.cn (T.D.)
2 College of Furnishings and Industrial Design, Nanjing Forestry University, Nanjing 210037, China
* Correspondence: yanxiaoxing@njfu.edu.cn

Abstract: In order to prepare chitosan-coated pomelo peel flavonoid microcapsules with antibacterial properties, chitosan was used as the wall material for the purpose of coating the core material, pomelo peel flavonoids. The pH of the microcapsule crosslinking reaction was 7.5, the mass ratio of the microcapsule core material to the wall material was 1:1, and the concentration of the emulsifier was 1%. The microcapsules obtained under these preparation conditions exhibited superior performance, morphology, and dispersion. Additionally, the yield and coating rates were recorded at 22% and 50%, respectively. To prepare the paint film, the microcapsules were added into the coatings at varying concentrations of 0%, 3.0%, 6.0%, 9.0%, 12.0%, and 15.0%. The antibacterial efficacy of the paint film for both bacteria was progressively enhanced with the incorporation of microcapsules. The antibacterial efficacy against *Staphylococcus aureus* was observed to be higher than that against *Escherichia coli*. As the content of microcapsules increased, the color difference in the paint film increased, the gloss loss rate increased, and the light transmission rate reduced. The tensile property and elongation at break reduced, and the roughness increased. At a microcapsule content of 6.0%, the paint film exhibited superior overall performance, with an antibacterial efficacy against *Escherichia coli* and *Staphylococcus aureus* of 46.3% and 56.7%, respectively. The color difference was 38.58. The gloss loss rate was 41.0%, the light transmission rate was 90.4%, and the paint film exhibited a large elastic region, with an elongation at break of 21.5% and a roughness of 1.46 μm.

Keywords: microcapsules; pomelo peel flavonoids; antibacterial properties; paint film

1. Introduction

With the innovation of technology, users expect more from furniture products [1–5]. As one of the commonly used materials in homes, wood materials have the characteristics of texture, easy to work with, light weight, and high strength [6–8]. However, wood has the property of dry shrinkage and wet expansion [9–11] and is vulnerable to damage such as decay and mold during use [12–14]. Wood absorbs moisture easily from the air, providing a favorable environment for bacteria to thrive [15]. Paint films are widely used in people's daily life in direct contact with the surface of household products. To protect the substrate of wooden furniture, the surface is usually coated [16–20]. The paint films on furniture surfaces are important carriers of bacterial transmission, so improving the antibacterial property of paint films is one of the effective means to stop the direct transmission of bacteria [21–24]. At present, plant-based antibacterial agents are the most extensively studied and conveniently extracted natural antibacterial agents. They use certain parts of plants containing bactericidal active substances or extract effective ingredients to achieve antibacterial effects. They have the advantages of high efficiency, low toxicity or nontoxicity, no pollution, and high selectivity and are not easily resistant to drugs [25–27]. Natural antibacterial agents have problems such as poor processing performance and unstable chemical properties [28]. Therefore, they cannot be directly applied in waterborne paint and need further research and promotion. Microcapsule technology is used to cover natural antibacterial or antibacterial substances to make antibacterial microcapsules [29,30], expanding the application scope of natural antibacterial agents [31,32].

Flavonoids are a class of compounds widely present in nature, found in certain plants and berries, with a total of about 4000 species. They often exist in the form of free or glycosidic compounds in plants and are currently a highly regarded natural active product with excellent physiological functions [33–35]. Lopes et al. [36] investigated the efficacy of glycosides and flavonoids in inhibiting *Staphylococcus aureus* RN4220 and *Staphylococcus aureus* SA1199B. The results showed that the tested flavonoids inhibited biofilm formation in *Staphylococcus aureus* strains. Liu et al. [37] studied the chemical composition of flavonoids extracted from pomelo peel. And its antibacterial activity against the selected pathogen *Vibrio anguillarum* and biosensor strain *Chromobacterium violaceum* CV026 was determined. These studies indicate that pomelo peel extract can be used for the production of antibacterial agents. Chitosan is a natural biopolymer which is widely used in various industries because of its excellent film forming property, antibacterial activity, and biodegradability [38,39]. Chen et al. [40] used chitosan as the wall material and water-soluble vitamin C and oil-soluble lemon essential oil as the core materials and prepared antibacterial microcapsules by the single condensation method. The results showed that both chitosan and oil-soluble lemon essential oil had antibacterial properties, so the paints with microcapsules had antibacterial properties. Utami et al. [41] encapsulated *Holothuria atra* microcapsules through the ionic gelation process of chitosan and sodium tripolyphosphate (STPP). The results showed that the prepared microcapsules had high antibacterial activity against *Escherichia coli* and *Staphylococcus aureus*. *Escherichia coli* and *Staphylococcus aureus* are commonly used for antibacterial performance testing. *Escherichia coli* is a Gram-negative bacterium, while *Staphylococcus aureus* is a Gram-positive bacterium, and the two bacteria can represent different bacterial species. *Escherichia coli* and *Staphylococcus aureus* are considered to be the two most common bacteria [42,43], which are widely found in nature and are closely related to human health.

Therefore, chitosan-coated pomelo peel flavonoid microcapsules were prepared with pomelo peel flavonoid as the core material, chitosan as the wall material, and STPP as the crosslinking agent. The microcapsules were prepared by the ion condensation method [44]. Different contents of microcapsules were added to the paint films to explore their application in the paint film. The performances of the paint film with added microcapsules were analyzed. The content of microcapsules in the paint film was controlled to make sure that the antibacterial properties of the paint film are improved while the film still has good properties. This provides a certain technical basis for the application of antibacterial microcapsules and the preparation of antibacterial paint film.

2. Materials and Methods

2.1. Materials

The deacetylation degree range of chitosan is 80.0% to 95.0%. A mold made of silicone material with a diameter of 50 mm × 50 mm × 10 mm was used to prepare the paint films. The waterborne acrylic topcoat used in the test was from Jiangsu Haidian Technology Co., Ltd., Jurong, China. The pomelos were Shatian Pomelo from Rong County, Yulin, China. Two microorganisms, *Staphylococcus aureus* and *Escherichia coli*, were used to test the antibacterial performance of the paint films. In addition, the materials used in this experiment were all chemicals. The experimental materials are shown in Table 1. The equipment used in the experiment is shown in Table 2.

Table 1. List of materials.

Material	Molecular Formula	CAS No.
Chitosan	$(C_6H_{11}NO_4)_n$	9012-76-4
Acetic acid	CH_3COOH	64-19-7
Tween-80	$C_{24}H_{44}O_6$	9005-65-6
Sodium tripolyphosphate	$Na_5P_3O_{10}$	7758-29-4
Anhydrous ethanol	C_2H_6O	64-17-5
Staphylococcus aureus	-	-
Escherichia coli	-	-
Nutrient agar medium	-	-
Nutritious broth	-	-

Table 2. Experimental equipment.

Equipment	Model	Manufacturing
Optical microscope (OM)	AX10	Carl Zeiss AG. Oberkochen, Germany
Scanning electron microscope (SEM)	Quanta-200	Thermo Fisher Scientific, Waltham, MA, USA
Magnetic stirrer	DF-101Z	Nanbei Scientific Instrument Technology Co., Ltd., Beijing, China
Fourier Transform Infrared (FTIR) spectrometer	VERTEX 80V	Bruker Corporation, Karlsruhe, Germany
Humidity chamber	HWS-50	Shanghai Shangyi Instrument Equipment Co., Ltd., Shanghai, China
Freeze-dryer	YTLG-10A	Shanghai Yetuo Technology Co., Ltd., Shanghai, China
Colony counter	XK97-A	Hangzhou Qiwei Instrument Co., Ltd., Hangzhou, China
Portable colorimeter	SC-10	Zhuhai Tianchuang Instrument Co., Ltd., Zhuhai, China
Gloss meter	HG268	Shenzhen ThreeNH Technology Co., Ltd., Shenzhen, China
Ultraviolet spectrophotometer	U-3900	Hitachi Scientific Instruments (Suzhou) Co., Ltd., Suzhou, China
Universal mechanical testing machine	AGS-X	Shimadzu Manufacturing Co., Ltd., Kyoto, Japan
Fine roughness tester	JB-4C	Suliang Instrument Technology Co., Ltd., Suzhou, China

2.2. Method of Preparing Microcapsules

As shown in Table 3, the optimal preparation process for the preparation of chitosan-coated pomelo peel flavonoid microcapsules was investigated by designing a three-factor two-level orthogonal test and a one-factor test. The orthogonal tests were designed, and the influencing factors were the pH for the crosslinking reaction of the microcapsules, the mass ratio of the core and wall materials of the microcapsules, and the concentration of the emulsifier during the emulsification of the microcapsules. Samples of microcapsules with different parameters were obtained. The experimental arrangement is shown in Table 4. According to the various test results, the maximum influence of the preparation of microcapsules was explored. The maximum influence factor was used as the variable of the one-factor test, and the preparation process of microcapsules was optimized through the test so as to obtain a better preparation process for the microcapsules.

Table 3. The orthogonal test schedule.

Level	Factor A pH Value	Factor B m(Core Material):m(Wall Material)	Factor C Concentration of Emulsifier (%)
1	6	1.0:1	1
2	8	1.2:1	3

Table 4. The orthogonal test microcapsule samples.

Sample	pH	m(Core Material):m(Wall Material)	Concentration of Emulsifier (%)
1	6	1.0:1	1
2	6	1.2:1	3
3	8	1.0:1	3
4	8	1.2:1	1

The preparation of microcapsules was divided into four steps. The first step was to prepare the chitosan wall material solution, the second step was to prepare the core material emulsion of pomelo peel flavonoids, the third step was to mix the wall material solution with the core material emulsion, and the fourth step was to add STPP into the mixed solution to make it crosslinked with the wall material. The specific preparation steps are shown in [45].

Table 5 shows the dosage of materials.

Table 5. The schedule of test material dosage.

Test	Sample	Chitosan (g)	1% Acetic Acid Solution (g)	Pomelo Peel Flavonoids (g)	Anhydrous Ethanol (g)	Emulsifier (g)	Deionized Water (mL)	STPP (g)
)rthogonal test	1	0.80	39.20	0.80	7.20	0.72	71.28	0.80
	2	0.80	39.20	0.96	8.64	2.11	68.29	0.80
	3	0.80	39.20	0.80	7.20	2.10	69.90	0.80
	4	0.80	39.20	0.96	8.64	0.70	69.70	0.80
One-factor test	5	0.80	39.20	0.80	7.20	0.72	71.28	0.80
	6	0.80	39.20	0.80	7.20	0.72	71.28	0.80
	7	0.80	39.20	0.80	7.20	0.72	71.28	0.80
	8	0.80	39.20	0.80	7.20	0.72	71.28	0.80
	9	0.80	39.20	0.80	7.20	0.72	71.28	0.80

2.3. *Preparation of Waterborne Paint Films with Different Contents of Microcapsules*

Waterborne paint films with different contents of microcapsules were coated in the silicone mold. The ordinary content of the surface paint of the substrate was 60 g/m^2–80 g/m^2, and a total of 4–6 coats were applied. To simulate the regular use of the paint on the surface of the substrate and to take into account the loss during actual use, the total waterborne paint mass of the paint film was set at 400 g/m^2, and the thickness was about 80 μm. The amount of paint film prepared is detailed in Table 6.

Table 6. The preparation of the paint film.

Content of Microcapsules (%)	Mass of Microcapsules (g)	Mass of Waterborne Paint (g)
0	0	1.00
3.0	0.03	0.97
6.0	0.06	0.94
9.0	0.09	0.91
12.0	0.12	0.88
15.0	0.15	0.85

The optimal microcapsule sample 7 prepared in the one-factor test was added to the waterborne paint at different concentrations of 0%, 3.0%, 6.0%, 9.0%, 12.0%, and 15.0%. The microcapsules and the waterborne paint were thoroughly stirred by a glass rod and coated in silicone molds. The silicone molds were left at room temperature for 1 h and then cured in an oven at 50 °C for 30 min. Finally, the paint film was slowly removed from the silicone mold to obtain a complete paint film sample for testing.

2.4. *Performance Test*

2.4.1. The Microcapsule Yield and Coverage Rate

After the preparation of microcapsules, the product quality was measured, and the percentage of the product quality to the raw material quality was calculated, that is, the yield of the microcapsules.

The weighed microcapsule powder, denoted as M_1, fully ground, was added to a beaker filled with anhydrous ethanol and soaked for 2 days. In order to fully dissolve the cores, the beakers were placed in a water bath at 60 °C for 4 h. After soaking, complete filtration, and rinsing, the resulting product was oven-dried at 50 °C until the weight did not change. The resulting product was recorded as M_2. This was the weight of the wall material of the microcapsule. The formula for calculating the coverage rate of microcapsules is shown in Formula (1).

$$P = (M_1 - M_2)/M_1 \times 100\% \quad (1)$$

2.4.2. The Morphology and Chemical Composition

An OM was used to observe and analyze the microstructure of the microcapsules. For observation, the light source was first turned on, and the slide was placed on the carrier table and fixed with a pressure clamp. The focus was adjusted until the field of view was clear, and the picture was taken.

An SEM was used to observe and analyze the microstructure of the microcapsules and the paint film. The SEM was used to characterize the microstructure of the paint films. In order to characterize the microcapsules by SEM, a processed sample was placed on the sample stage, a door of the sample compartment was closed, and the vacuum pump was turned on to pump the air. The relevant parameters were set according to the sample characteristics and observation need, and the desired image was captured and saved. When the particle size of the microcapsules was analyzed, the SEM image of the microcapsules obtained was analyzed by software to generate a histogram of the particle size of the microcapsules.

An FTIR spectrometer was used to determine the chemical composition of microcapsules and the paint films with added microcapsules. When the paint film samples were tested by the FTIR spectrometer, the parameters were adjusted so that the background spectra were first acquired, and subsequently, the paint film samples were placed in the sample chamber for testing. The KBr pressing method was used for the testing of microcapsule samples.

2.4.3. The Antibacterial Performance

Escherichia coli (ATCC25922) and *Staphylococcus aureus* (ACTT6538) were used to test the antibacterial properties of the coating. According to GB/T 21866-2008 [46], the testing was divided into bacterial strain preservation, bacterial strain activation, bacterial suspension preparation, sample testing, and calculation results. The specific experimental steps can be found in [46,47]. The antibacterial rate R of the paint film was calculated according to Formula (2), where B represents the average number of recovered colonies after 48 h of blank coating and C represents the average number of recovered colonies after 48 h of antibacterial paint film, in CFU/piece.

$$R = (B - C)/B \times 100\% \quad (2)$$

2.4.4. The Optical Performance

Based on the GB/T 11186.3-1989 [48], a portable colorimeter was used to measure the color value of the paint film and calculate the color difference in the paint film. After the portable colorimeter was calibrated, a point was selected on the paint film to test, and the color values of L, a, and b were noted. The values of the blank paint film were denoted as L_1, a_1, and b_1, and those of the film containing microcapsules were denoted as L_2, a_2, and b_2. The color difference ΔE of the paint film was calculated by Formula (3), where $\Delta L = L_2 - L_1$, $\Delta a = a_2 - a_1$, and $\Delta b = b_2 - b_1$.

$$\Delta E = \left[(\Delta L)^2 + (\Delta a)^2 + (\Delta b)^2\right]^{1/2} \quad (3)$$

Based on GB/T 4893.6-2013 [49], a gloss meter was used to test the glossiness of the paint film and record the glossiness of the paint film under three sets of incident angles (20°, 60°, and 85°). G_0 was the glossiness of the paint film without microcapsules added, G_1 was the glossiness of the paint film containing microcapsules, and the gloss loss rate G_L

at 60° incidence angle of the paint film was calculated. The calculation formula is shown as Formula (4).

$$G_L = (G_0 - G_1)/G_0 \times 100\% \quad (4)$$

An ultraviolet spectrophotometer was used to measure the light transmittance of the paint film in the visible band with a wavelength range of 380 nm–780 nm.

2.4.5. The Mechanical Performance

The paint film was tested by a universal mechanical testing machine. During the test, the two ends of the paint film were fixed by the fixture, and the paint film was stretched until it broke. According to Formula (5), the elongation at break of the film was denoted as e, and the original length of the film was denoted as L_0. When the film broke, the distance beyond the original length was denoted as L_n. The elongation at break of the paint film was the ratio of the distance over the original length to the original length.

$$e = L_n/L_0 \times 100\% \quad (5)$$

A fine roughness tester was used to test the roughness of the paint film.

3. Results and Discussion

3.1. The Microcapsules' Yield and Coverage Rate

Four kinds of microcapsule samples were obtained by a three-factor and two-level orthogonal test. The yield of the orthogonal test for the preparation of microcapsules is shown in Table 7. Sample 3 has the highest yield, and the result is 22%. According to Table 7, it can be found that the yield results of microcapsule samples have little difference. According to the yield range analysis, the primary and secondary effects of the three factors are A > B = C. Therefore, the biggest factor that influences the microcapsule yield is A, which is the pH value of the microcapsule crosslinking reaction. The mass ratio of the microcapsule core material to wall material and the emulsifier concentration had the same effect on the yield. The optimal level of microcapsule yield was A2 B1 C2, namely, a crosslinking reaction pH of 8, a core–wall ratio of 1:1, and an emulsifier concentration of 3%. As shown by the result data in Table 7, the size of factors affecting the yield of microcapsules is basically consistent with the range result. Through the sum of squared deviations, it can be seen that factor A has a much greater impact on the yield than the other two factors. Therefore, combining the yield results of microcapsules, it can be concluded that the pH value of the microcapsule crosslinking reaction has the greatest influence on the yield of the microcapsules. The optimum preparation process of the microcapsules was when the pH value of the microcapsule crosslinking reaction was 8, the mass ratio of the microcapsule core material to wall material was 1:1, and the emulsifier concentration was 3%.

Table 7. Analysis of range and variance results of microcapsule yield rate in orthogonal tests.

Category	Sample	Factor A pH Value	Factor B m(Core Material):m(Wall Material)	Factor C Concentration of Emulsifier (%)	Yield (%)
Range	1	6	1.0:1	1	20
	2	6	1.2:1	3	20
	3	8	1.0:1	3	22
	4	8	1.2:1	1	21
	Mean value 1	20.000	21.000	20.500	
	Mean value 2	21.500	20.500	21.000	
	R	1.500	0.500	0.500	
	Factor primary and secondary levels		A > B = C		
	Optimal level	A2	B1	C2	
	Optimal solution		A2 B1 C2		
Variance	Sum of squared deviations	2.250	0.250	0.250	
	Degree of freedom	1	1	1	
	F_{ratio}	2.455	0.273	0.273	
	$F_{critical\ value}$	10.100	10.100	10.100	
	Significance				

Table 8 shows the coverage rate results of the chitosan-coated pomelo peel flavonoid microcapsules in the orthogonal test. Sample 3 had the highest coverage rate of 53%. According to the analysis of range results, the order of the three influencing factors is A > B > C. Therefore, the factor that has the largest influence on microcapsule coverage rate is the pH value of the crosslinking reaction, followed by the mass ratio of the microcapsule core–wall material and finally the emulsifier concentration in the preparation. A2 B1 C1 is the optimal coverage rate level of the microcapsule preparation. The variance results of the impact of various factors on the coverage rate are basically consistent with the range results. From the sum of squares data, it can be seen that the influence of the pH value of the microcapsule crosslinking reaction on the result of the coverage rate is much greater than that of others. Thus, it can be concluded that factor A, that is, the pH value in the microcapsule crosslinking reaction, has the biggest influence on the coverage rate of microcapsules. The optimal process for the coverage rate of microcapsules is the microcapsule crosslinking reaction with a pH value of 8, a core–wall ratio of 1:1, and a concentration of the emulsifier of 1%.

Table 8. Analysis of range and variance results of microcapsule coverage rate in orthogonal tests.

Category	Sample	Factor A pH Value	Factor B m(Core Material):m(Wall Material)	Factor C Concentration of Emulsifier (%)	Coverage Rate (%)
Range	1	6	1.0:1	1	45
	2	6	1.2:1	3	31
	3	8	1.0:1	3	53
	4	8	1.2:1	1	51
	Mean value 1	38.000	49.000	48.000	
	Mean value 2	52.000	41.000	42.000	
	R	14.000	8.000	6.000	
	Factor primary and secondary levels		A > B > C		
	Optimal level	A2	B1	C1	
	Optimal solution		A2 B1 C1		
Variance	Sum of squared deviations	196.000	64.000	36.000	
	Degree of freedom	1	1	1	
	F_{ratio}	1.986	0.649	0.365	
	$F_{critical\ value}$	10.100	10.100	10.100	
	Significance				

According to the results of the orthogonal test, the pH value of the microcapsule crosslinking reaction is the most significant factor influencing the yield and coverage rate of chitosan-coated pomelo peel flavonoid microcapsules. Therefore, based on the orthogonal test, a one-factor test was designed with the pH value of the microcapsule crosslinking reaction as the variable. According to the orthogonal test results, the yield of microcapsule samples did not vary much, so the preparation parameters of the one-factor test focused on the results of the coverage rate in the orthogonal test. When the mass ratio of the core–wall material is 1:1, the coverage rate and yield are better. When the emulsifier concentration is 1%, the coverage rate is better, and the high coverage rate is conducive to improving the antibacterial property of the waterborne paint, so the emulsifier concentration is determined to be 1% in the one-factor test. The results of the one-factor test are shown in Table 9. In the one-factor test, sample 5 with a pH value of 5.5 had a higher yield but a lower coverage rate. This may be because the pH value was too low, resulting in a large amount of residual chitosan wall material being coated. It can be proved that the coating of microcapsules should be carried out under alkaline conditions. When the pH exceeds 5.5, the yield of microcapsules slightly increases with the increase in pH value. When the pH value reaches 7.5, the coverage rate of microcapsules is up to 50%, and with the increase in pH value, the coverage rate of the microcapsules gradually decreases, which may be because the high pH value leads to the formation of microcapsules too fast, and the core material is reduced. It is proved that when pH value is 7.5, it is more suitable for pomelo peel flavonoids coated with chitosan. The optimal preparation conditions are a core-to-wall ratio of 1:1, an emulsifier concentration of 1%, and a pH value of 7.5 during the crosslinking reaction. Based on the results of the microcapsules' yield, coverage rate, and morphology in the one-factor test, sample 7 with a pH value of 7.5 at the time of the crosslinking reaction was selected as the optimal sample and added to the waterborne paint film to further explore the application of microcapsules in the waterborne paint.

Table 9. The yield and coverage rate of microcapsules in the one-factor test.

Sample	pH Value	Yield Rate (%)	Coverage Rate (%)
5	5.5	24	38
6	6.5	20	41
7	7.5	22	50
8	8.5	22	47
9	9.5	23	44

3.2. The Morphological and Chemical Composition of the Microcapsules

The macroscopic morphology of the pomelo peel flavonoids, chitosan, and chitosan-coated pomelo peel flavonoid microcapsule sample 7 is shown in Figure 1. The pomelo peel flavonoids are irregular large granular yellow crystals, and the chitosan is a beige powder. The microcapsules prepared with these two are in a yellow powder shape. This is because during the preparation of the microcapsules, the core emulsion mixed with the wall solution, resulting in staining. At the same time, the chitosan wall material itself is thinner and has a beige color, and there are also more encapsulated core materials. The color of the core materials is revealed through more micropores, making the color of the microcapsules darker. This can preliminarily prove that the successfully prepared microcapsules contain two substances, pomelo peel flavonoids and chitosan.

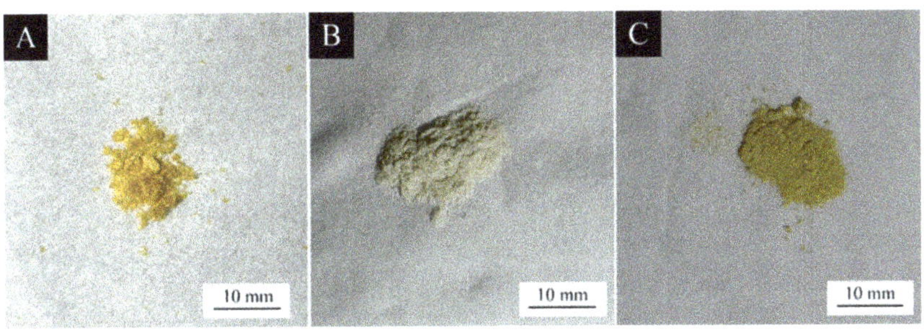

Figure 1. Macroscopic morphology: (**A**) pomelo peel flavonoids for core material, (**B**) chitosan for wall material, and (**C**) microcapsules (sample 7).

The micromorphology of the chitosan-coated pomelo peel flavonoid microcapsules prepared by orthogonal test is shown in Figure 2. Microscope images show that there are two kinds of materials in the microcapsule: the wall material, chitosan, and the core material, pomelo peel flavonoids. The bright white spot inside the microcapsule is the pomelo peel flavonoids, which can preliminarily prove the success of the microcapsule coating. It can be observed that microcapsule sample 1 and sample 2 prepared at pH 6 in the orthogonal test have better dispersibility and more microcapsule particles. The agglomeration of microcapsule sample 3 and sample 4 was serious when the pH value was 8. This may be due to the high pH value during the reaction of microcapsules, which leads to a fast crosslinking reaction of the microcapsules. When the core material is not fully coated, too much crosslinking is formed in the wall material; the microcapsules were adhered together by a large number of walls that were not coated with the core material and could not be presented as separate spheres.

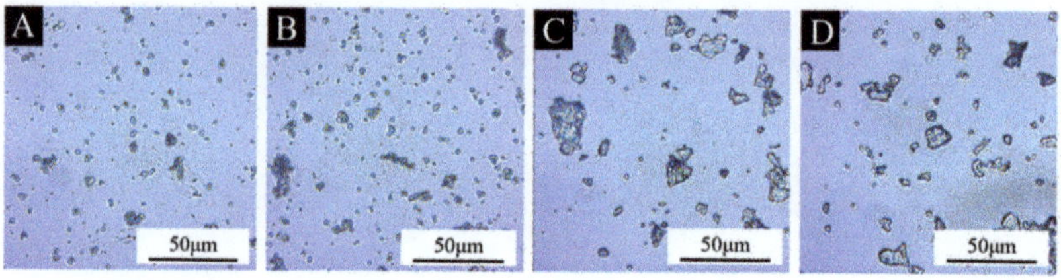

Figure 2. OM images of the microcapsules: (**A–D**) samples 1–4.

Figure 3 shows the microscopic morphology of the chitosan-coated pomelo peel flavonoid microcapsules in the one-factor test. The particles of chitosan in the microscope image are too small, and the observation analysis is limited. Therefore, the microscopic morphology of microcapsules in the one-factor test is mainly observed by the SEM. It can be observed that microcapsule sample 5 in the one-factor test is basically in an amorphous state, which indicates that the crosslinking reaction of microcapsules is not suitable when the pH is 5.5, and microcapsules cannot be formed when the pH is too low. The agglomeration phenomenon of microcapsule sample 8 was more serious, and the agglomeration phenomenon of microcapsule sample 9 was the most serious, but it was obvious that the microcapsules had formed into a spherical state, indicating that the microcapsules were successfully prepared. However, a high pH value will lead to a sticky agglomeration of microcapsules, which is not conducive to microcapsule dispersion and is not suitable for the microcapsule crosslinking reaction. The crosslinking reaction between the chitosan and STPP occurred under alkaline conditions. When the solution is at a lower pH value, STPP is more likely to protonate, forming positively charged STPP molecules. The positively charged STPP interacts electrostatically with the negatively charged chitosan molecules, promoting the crosslinking reaction. When the pH value is higher, the protonation degree of STPP decreases, and the electrostatic interaction between STPP and chitosan is weakened, so the crosslinking reaction is weakened. In Figure 3, the microcapsules prepared at pH 7.5 for sample 7 were clearly observed to be more complete spherical microcapsules and well dispersed, with fewer impurities compared to the other samples. Therefore, considering the yield, coverage rate, and microscopic morphology of the chitosan-coated pomelo peel flavonoid microcapsules, sample 7 with a 22% yield and a 50% coverage rate was selected. Figure 4 shows the particle size distribution of microcapsules prepared by the one-factor test. As shown in Figure 4, the particle size of the microcapsules prepared at different pH values is relatively dispersed, mainly between 2 and 7 μm, indicating that the chitosan-coated pomelo peel flavonoid microcapsules have different particle sizes and poor morphology. The average particle size of the four samples is not significantly different. The average particle size of sample 6 is 5.19 μm, that of sample 7 is 5.19 μm, that of sample 8 is 5.28 μm, and that of sample 9 is 5.38 μm.

Figure 5 shows the infrared spectra of the wall material, the chitosan-coated pomelo peel flavonoid microcapsules, and the core material. In the infrared spectra of the chitosan wall material, 3433 cm^{-1} is a vibration absorption peak of the chitosan structure -OH, 1647 cm^{-1} is the vibration absorption peak of an amide group I, 1088 cm^{-1} is a C-O-C absorption peak, and 891 cm^{-1} is the stretching vibration absorption peak of a pyranoside ring. In an infrared spectrum of the pomelo peel flavonoids, the vibration peaks at 1372, 1242, and 1058 cm^{-1} are the characteristic peaks of pomelo peel flavonoids, the absorption peak at about 3390 cm^{-1} is a hydroxyl association, and the absorption peak at 2932 cm^{-1} is CH_2-. In addition, 1740 cm^{-1} and 1646 cm^{-1} are the C-O-induced stretching vibration peaks, and 877 cm^{-1} is the absorption peak caused by substituent positions on the benzene ring. The absorption curves also showed the absorption peaks of the chitosan and pomelo peel flavonoids [50]. The chemical components of the chitosan and pomelo peel flavonoids were present in the microcapsules, which proved the successful preparation of microcapsules.

Figure 3. SEM images of one-factor microcapsules: (**A–E**) samples 5–9.

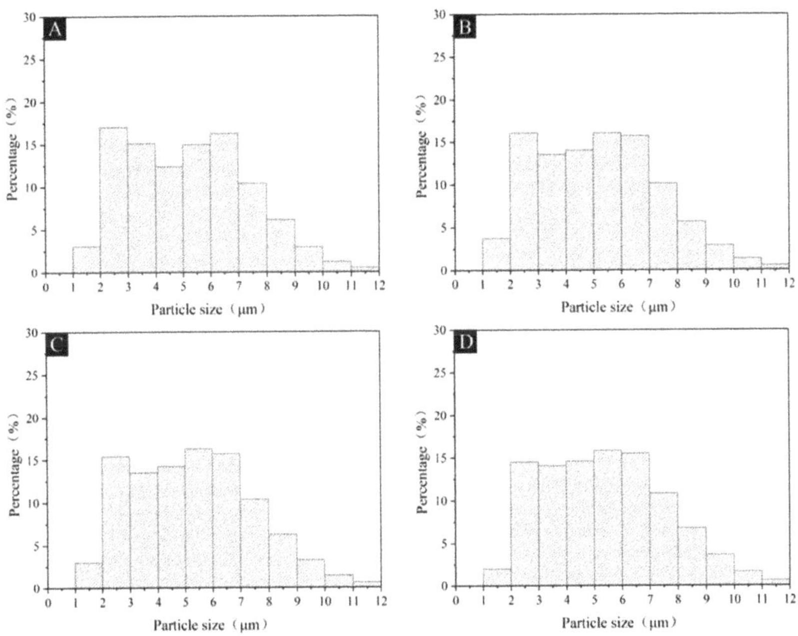

Figure 4. Particle size distribution: (**A–D**) sample 6–9.

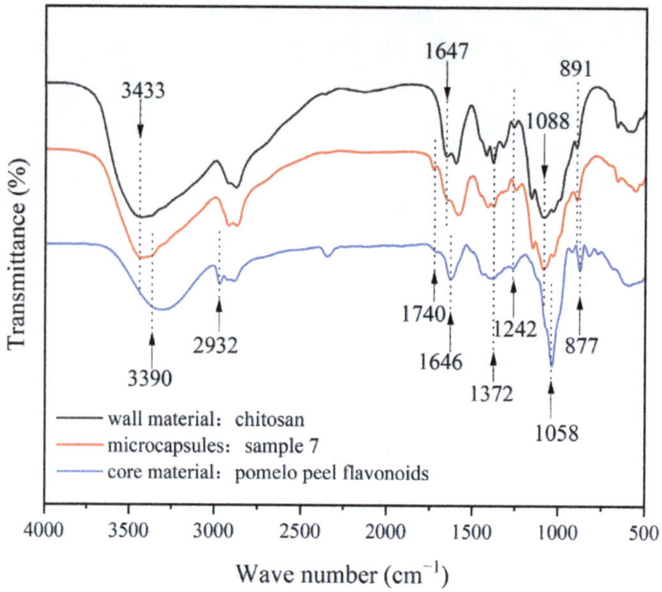

Figure 5. FTIR images of chitosan, pomelo peel flavonoids, and microcapsule sample 7.

3.3. *The Morphology of the Paint Film*

The macroscopic morphology shown in Figure 6 is the paint film with different contents of sample 7. The paint film was yellow and transparent after microcapsules were added. With the increase in microcapsule content in the paint film, the transparency gradually decreased, and the color of the paint film gradually deepened. This is because the microcapsules are yellow; the addition of the waterborne paint changes the color of the film. The microcapsules coated with chitosan have good dispersibility and no obvious granular sensation in the waterborne paint. When the proportion of microcapsule mass in the total paint film mass is less than 15.0%, the morphology of the paint film is not affected except for the color change, and no cracking phenomenon appears, which preliminarily proves that the chitosan-coated pomelo peel flavonoid microcapsules are suitable for addition to the waterborne paint. When the proportion of microcapsule mass in the total paint film mass reaches 15.0%, the edges of the paint film have already been damaged. This is because the chitosan increases the viscosity of the paint film. The excessive content of microcapsules in the waterborne paint leads to increased viscosity and poor flat property, making it difficult to form the paint film. This indicates that the paint film with 15.0% microcapsules is no longer suitable for application.

The microstructure of the paint film with added microcapsule sample 7 is shown in Figure 7. With the addition of 3.0% microcapsules, the paint film has a slightly raised granular feeling. When the content is 6.0% and 9.0%, the granular feeling on the paint film surface is more obvious. When the content reaches 12.0%, the agglomeration of a lot of microcapsules in the paint film, the paint film presents a large area of raised folds. It can be seen that when the content of microcapsules is above 9.0%, it will seriously affect the flatness.

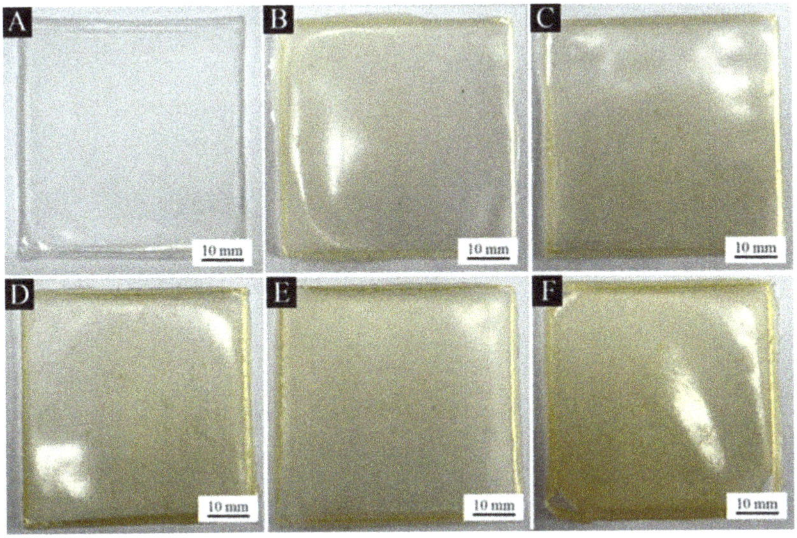

Figure 6. Macroscopic morphology of the paint films with different contents of microcapsules: (**A**) without microcapsules, (**B**) with 3.0% microcapsules, (**C**) with 6.0% microcapsules, (**D**) with 9.0% microcapsules, (**E**) with 12.0% microcapsules, and (**F**) with 15.0% microcapsules.

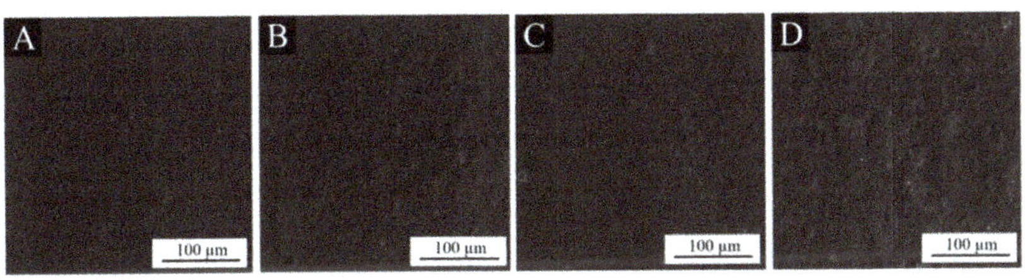

Figure 7. SEM images of paint films with different microcapsule contents: (**A**) with 3.0% microcapsules, (**B**) with 6.0% microcapsules, (**C**) with 9.0% microcapsules, and (**D**) with 12.0% microcapsules.

3.4. The Chemical Composition of the Paint Film

Figure 8 shows the FTIR image of the paint film without microcapsules and added microcapsules with a content of 9.0%. In Figure 8, around 1726 cm^{-1} is the characteristic peak of the C=O in the waterborne acrylic resin in the waterborne acrylic paint, and around 1144 cm^{-1} is the vibration peak of the C-O in both the waterborne paint and microcapsules. About 3390 cm^{-1} is the characteristic peak of the hydroxy-association of pomelo peel flavonoids in microcapsules, and 2932 cm^{-1} is the absorption peak of CH_2-. The C-O-C absorption peak of the chitosan structure in the microcapsules is about 1088 cm^{-1}, and the stretching vibration absorption peak of the pyranoside ring is at 891 cm^{-1}. It can be proved that the microcapsules are used in the waterborne paint, and the chemical composition of the paint film is not destroyed.

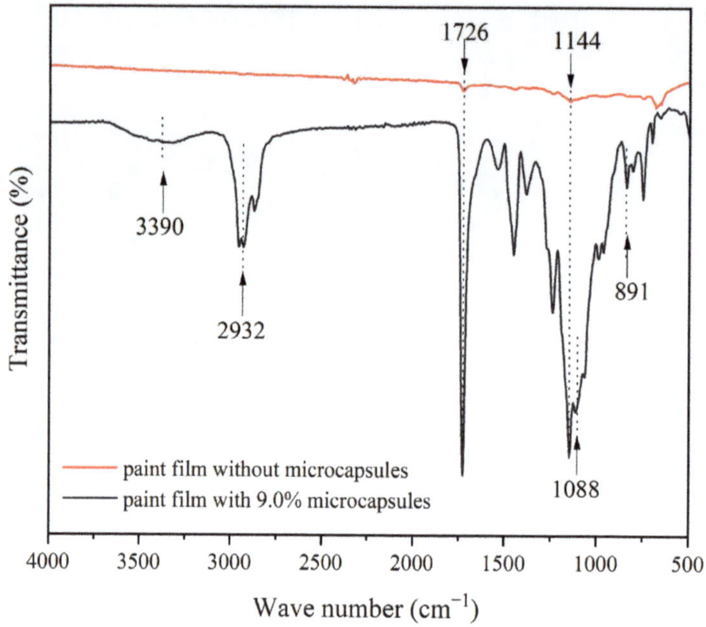

Figure 8. FTIR image of the paint films with and without microcapsules.

3.5. The Antibacterial Properties of the Paint Film

The average number of recovered bacteria and the antibacterial rate are shown in Table 10. As shown in Table 10, the antibacterial rate against both bacteria gradually increased with the increase in the microcapsule content, indicating that the paint film with added microcapsules has an antibacterial effect on both bacteria. When the content of microcapsules in the paint film reaches 15%, the antibacterial rate of the paint film against the two kinds of bacteria could reach 83.2% and 91.7%, respectively. Both the wall material chitosan and the core material pomelo peel flavonoids have antibacterial activity, which further improves the antibacterial rate of the microcapsules.

Table 10. The antibacterial properties.

Content of Microcapsules (%)	Average Number of Recovered *Escherichia coli* (CFU·Piece^{-1})	Antibacterial Rate against *Escherichia coli* (%)	Average Number of Recovered *Staphylococcus aureus* (CFU·Piece^{-1})	Antibacterial Rate against *Staphylococcus aureus* (%)
0	190	-	432	-
3.0	134	29.5 ± 0.6	289	33.1 ± 0.8
6.0	102	46.3 ± 1.6	187	56.7 ± 1.5
9.0	73	61.6 ± 0.9	126	70.8 ± 1.2
12.0	51	73.2 ± 1.4	74	82.9 ± 2.7
15.0	32	83.2 ± 1.1	36	91.7 ± 1.7

Figure 9 shows the antibacterial rate of paint films with different microcapsule contents. The antibacterial activity of the paint film against *Staphylococcus aureus* is slightly higher than that against *Escherichia coli*. The results show that the chitosan-coated pomelo peel flavonoid microcapsules successfully had an antibacterial effect in the waterborne paint film and effectively improved the difficult mixing of the pomelo peel flavonoids in waterborne paint. The pomelo peel flavonoids that have antibacterial properties work together with the chitosan that also has antibacterial properties through the micropores of the microcapsules to inhibit bacteria.

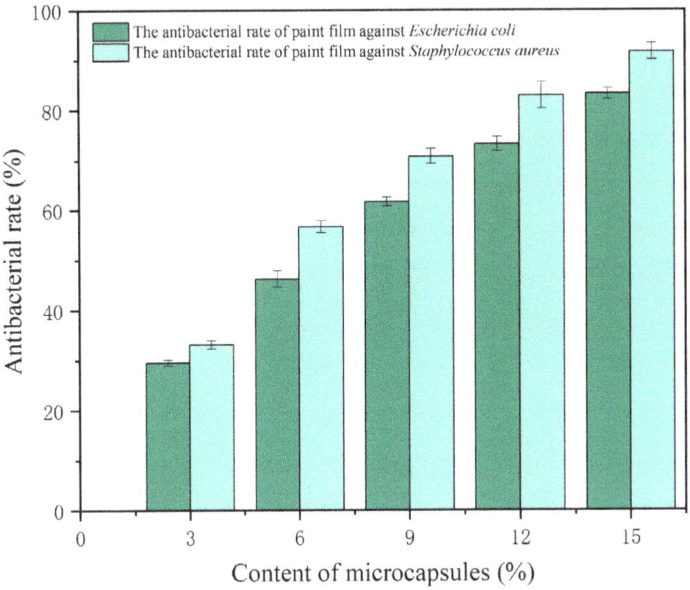

Figure 9. The antibacterial rate of paint films with different microcapsule contents.

The chitosan-coated pomelo peel flavonoid microcapsules had a higher antibacterial rate compared to melamine-resin-coated pomelo peel flavonoid microcapsules [51]. When the content of melamine-resin-coated pomelo peel flavonoid microcapsules in the paint film was 6%, the antibacterial rate against *Escherichia coli* was 50.5%, and the antimicrobial rate against *Staphylococcus aureus* was 40.5%. When the content of the chitosan-coated pomelo peel flavonoid microcapsules in the paint film was 6%, the antibacterial rates against *Escherichia coli* and *Staphylococcus aureus* were 46.3% and 56.7%, respectively. This is due to the fact that only the core material is antibacterial in the melamine-resin-coated pomelo peel flavonoid microcapsules, and the wall material is not antibacterial, whereas chitosan, which is used as the wall material of the microcapsules in the chitosan-coated pomelo peel flavonoid microcapsules, is also antibacterial, so the antibacterial rate of the microcapsules is increased.

A significance analysis was conducted on the data in Table 10. The significance analysis was conducted using a non-repeated two-factor variance method. The three values of F, p-value, and F crit separately were calculated (Table 11). Among them, F is the test statistic, a statistical measure used for hypothesis testing calculations. The p-value represents a significance level, evaluates the range and interval of the overall parameters, and calculates the probability of the experiment occurring. F crit is the critical value of F on the corresponding significance level. Among them, F > F crit indicates that there is a difference between the two sets of data. F < F crit means there is no difference between the two sets of data. The criteria for determining significant differences are $0.01 < p\text{-value} < 0.05$, indicating significant differences. A p-value ≤ 0.01 means the difference is extremely significant. A p-value > 0.05 indicates insignificance. According to the above methods, the results obtained are F > F crit and p-value < 0.01, indicating that the content of microcapsules in the paint film has a significant effect on the antibacterial properties of the paint film.

Table 11. Significance analysis of antibacterial properties.

Item	SS	d_f	MS	F	p-Value	F Crit
Content of microcapsules	10,537.29	5	2107.457	243.8058	0.00	5.050329
Antibacterial species	142.83	1	142.83	16.5236	0.009682	6.607891
Error	43.22	5	8.644			
Total	10,723.34	11				

3.6. The Optical Properties of the Paint Film

(1) The chromaticity value and color difference

The chromaticity value of and color difference in paint films with different microcapsule contents is shown in Table 12. When the paint film contains 15% microcapsules, the color difference can reach up to 40.74. Because of the addition of yellow granular microcapsules to the paint, the color of the paint film deepens, and the *L* value in the paint film gradually decreases as the microcapsules increase. The *a* values representing the red–green values are positive in all samples, indicating that the color was reddish. Because the color of the paint film itself tends to be blue, the *b* value of the sample without microcapsules is negative. After adding microcapsules, the *b* value changed from negative to positive, indicating that the color became yellowish and gradually increased as the microcapsules increased, indicating that the yellow color of the microcapsules themselves had an obvious influence on the paint film.

Table 12. The chromaticity value of and color difference in paint films with different microcapsule contents.

Content of Microcapsules (%)	*L*	*a*	*b*	ΔE
0	81.91	1.73	−2.27	-
3.0	44.93	0.97	0.50	37.09
6.0	43.50	0.97	1.30	38.58
9.0	43.00	0.90	1.57	39.11
12.0	42.20	0.87	1.60	39.91
15.0	41.40	0.63	1.87	40.74

(2) Glossiness and gloss loss rate

Table 13 shows the glossiness and gloss loss rate of paint films with different microcapsule contents. At different incidence angles, the glossiness decreases as the microcapsules increase. Because the chitosan-coated pomelo peel flavonoid microcapsules have color, the addition of microcapsules enhances the diffuse reflection of light on the surface of the paint film, which affects the glossiness. When the paint film contained 9.0% microcapsules, the gloss loss rate of the paint film was 41.7%; when the microcapsule content was higher than 9.0%, the gloss loss rate of the paint film began to decrease as the microcapsules increased. This is because of the good dispersibility of the chitosan-coated pomelo peel flavonoid microcapsules in the waterborne paint, which has little effect on the flatness of the dried paint film.

Table 13. The glossiness and gloss loss rate of paint films with different microcapsule contents.

Content of Microcapsules (%)	20° (%)	60° (%)	85° (%)	Gloss Loss Rate (%)
0	6.10	17.45	31.17	-
3.0	4.47	14.13	14.93	19.1
6.0	2.40	10.30	7.27	41.0
9.0	2.13	10.17	4.10	41.7
12.0	2.10	10.17	3.40	41.7
15.0	1.77	8.10	3.27	53.6

(3) Visible light transmittance

The visible light transmittance of paint films with different microcapsule contents is shown in Table 14. The visible light transmittance is higher than 85% on the whole. The addition of microcapsules has little effect on the visible light transmittance. When the paint film contained 3.0% microcapsules, the visible light transmittance value was 93.25%. Figure 10 shows the light transmittance curve of microcapsule paint film with different additive content. As the microcapsule content increases, the light transmittance of the paint film decreases slightly. All paint films tend to be relatively transparent, exhibiting a stable visible light wavelength range. This indicates that adding microcapsules has little effect on the transparency of the paint film and is suitable for practical applications.

Table 14. The visible light transmittance of paint films with different microcapsule contents.

Content of Microcapsules (%)	Visible Light Transmittance (%)
0	89.89
3.0	93.25
6.0	90.40
9.0	90.47
12.0	87.94
15.0	85.54

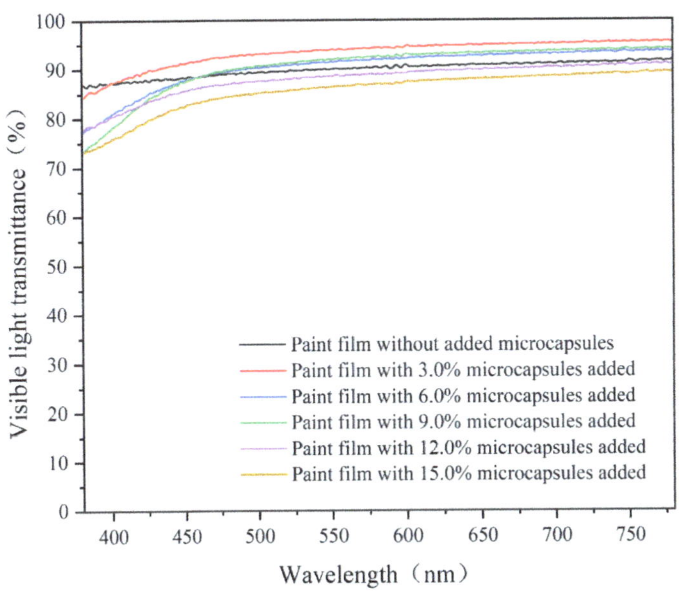

Figure 10. Visible light transmittance.

3.7. The Mechanical Properties of the Paint Film

(1) Roughness

The effect of microcapsule content on the roughness of the paint film is shown in Table 15. The roughness increased with the increase in microcapsules. The addition of microcapsules results in uneven particles on the paint film surface. When the microcapsule content increased, the concave and convex feeling of the paint film was enhanced, which affects the flatness of the paint film and makes the roughness rise continuously. However, because the microcapsules prepared with chitosan have better dispersibility in the waterborne paints, the overall roughness of the paint film is low.

Table 15. The effect of different microcapsule content on the roughness.

Content of Microcapsules (%)	Roughness (%)
0	0.27
3.0	0.46
6.0	1.46
9.0	2.60
12.0	2.65
15.0	2.70

(2) Elongation at break

As an important parameter characterizing the uniform or stable deformation of a material, elongation at break is an index used to judge the ductility of a material, indicating the ability of the paint film to deform without rupturing when subjected to an external force. Higher elongation at break indicates better ductility of the paint. The tensile properties of the paint film are shown in Figure 11. When the microcapsule content in the paint film was 3.0% and 6.0%, the elastic region of the paint film was the largest, and the ductility of the paint film was higher. This means that the paint film has better flexibility and ductility and can better adapt to the deformation of the substrate, thereby improving the overall performance of the paint. But when the microcapsule content exceeded 9.0%, the ductility of the paint film began to decrease gradually. This shows that the appropriate amount of microcapsules can enhance the ductility of the film, but this may be because the chitosan-coated pomelo peel flavonoid microcapsules dispersed in waterborne paints have a certain viscosity, which thickens the paint and regulates the viscosity of waterborne paints to a certain extent. However, the microcapsule content is too high, which will reduce the ductility of the paint film. Too many particles of microcapsules added to the paint film can easily produce agglomeration; microcapsules cannot be evenly distributed in the paint film, resulting in stress concentration, and these stresses will be released when the paint film is subjected to external forces, thus reducing the elongation at break.

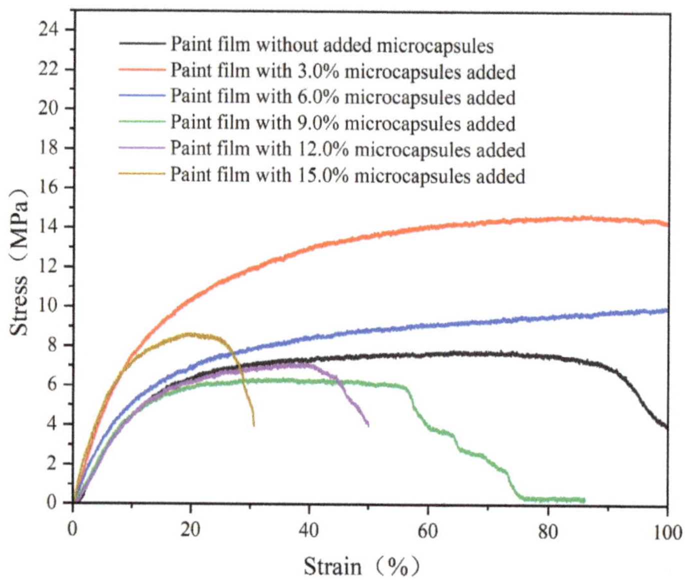

Figure 11. The tensile properties.

Table 16 shows the effect of the microcapsule content on the elongation at break. The elongation at break was 18.9% when the paint film did not contain microcapsules. When the content of microcapsules added to the paint film was 3%, the elongation at break was 22.3%. The chitosan-coated pomelo peel flavonoid microcapsules have a certain viscosity, and an appropriate amount of dispersion in the waterborne paints can thicken the paint and improve the viscoelasticity. The elongation at break of the membrane decreased when the content of microcapsules in the paint film was greater than 3.0%. When the content was 15.0%, the elongation at break was 7.9%. When the microcapsule content is too high, the viscoelasticity decreases, the paint film becomes brittle, and the relative tensile property of the paint film gradually decreases, which leads to a decrease in the elongation at break.

Table 16. The elongation at break.

Content of Microcapsules (%)	Elongation at Break (%)
0	18.9
3.0	22.3
6.0	21.5
9.0	16.1
12.0	10.5
15.0	7.9

4. Conclusions

The preparation procedure for the optimal chitosan-coated pomelo peel flavonoid microcapsules is as follows: the pH value of the microcapsule crosslinking reaction is 7.5, the core–wall ratio is 1:1, and the concentration of the emulsifier is 1%. The morphology of microcapsules is better, and the dispersion is relatively uniform. With the increase in microcapsule content, the antibacterial rate of the paint film with added microcapsules against the two kinds of bacteria increased gradually, and the antibacterial rate against *Staphylococcus aureus* was higher than that against *Escherichia coli*. The color difference in the paint film gradually increased, the gloss gradually decreased, the loss of light increased, the roughness gradually increased, and the light transmittance decreased. The ductility of the paint film could be enhanced with the appropriate amount of microcapsules. The best tensile properties and the highest elongation at break of 22.3% were obtained when the paint film contained 3.0% microcapsules. When the content of microcapsules exceeds 6.0%, the tensile performance and elongation at break of the paint film significantly decrease. When the content of microcapsules was 6.0%, the comprehensive performance of the paint film was superior, the antibacterial rate of the paint film against *Escherichia coli* was 46.3%, the antibacterial rate of the paint film against *Staphylococcus aureus* was 56.7%, the color difference was 38.58, the gloss loss rate was 41.0%, the transmittance rate was 90.4%, and the paint film had a large elastic area, with an elongation at break of 21.5% and a roughness of 1.46 μm. These results provide a technical reference for the application of pomelo peel flavonoid microcapsules in the preparation of waterborne antibacterial paints.

Author Contributions: Conceptualization, methodology, validation, resources, data management, and supervision, J.D.; writing—review and editing, T.D.; formal analysis and investigation, X.Y. All authors have read and agreed to the published version of the manuscript.

Funding: This project was partly supported by the Postgraduate Research and Practice Innovation Program of Jiangsu Province (SJCX24_0399) and the Natural Science Foundation of Jiangsu Province (BK20201386).

Institutional Review Board Statement: Not applicable.

Informed Consent Statement: Not applicable.

Data Availability Statement: Data are contained within this article.

Conflicts of Interest: The authors declare that no conflicts of interest.

References

1. Hu, J.; Liu, Y.; Xu, W. Impact of cellular structure on the thickness and light reflection properties of structural color layers o diverse wood surfaces. *Wood Mater. Sci. Eng.* **2024**. [CrossRef]
2. Wu, S.S.; Zhou, J.C.; Xu, W. A convenient approach to manufacturing lightweight and high-sound-insulation plywood usi furfuryl alcohol/multilayer graphene oxide as a shielding layer. *Wood Mater. Sci. Eng.* **2024**. [CrossRef]
3. Hu, W.; Yu, R. Study on the strength mechanism of the wooden round-end mortise-and-tenon joint using the digital ima correlation method. *Holzforschung* **2024**. [CrossRef]
4. Wang, C.; Zhang, C.Y.; Zhu, Y. Reverse design and additive manufacturing of furniture protective foot covers. *BioResources* **202** *19*, 4670–4678. [CrossRef]
5. Wang, C.; Yu, J.H.; Jiang, M.H.; Li, J.Y. Effect of selective enhancement on the bending performance of fused deposition metho 3D-printed PLA models. *BioResources* **2024**, *19*, 2660–2669. [CrossRef]
6. Zhang, N.; Xu, W.; Tan, Y. Multi-attribute hierarchical clustering for product family division of customized wooden doo *Bioresources* **2023**, *18*, 7889–7904. [CrossRef]
7. Hu, J.; Liu, Y.; Xu, W. Influence of cell characteristics on the construction of structural color layers on wood surfaces. *Forests* **202** *15*, 676. [CrossRef]
8. Mai, C.; Schmitt, U.; Niemz, P. A brief overview on the development of wood research. *Holzforschung* **2022**, *76*, 102–119. [CrossRe
9. Xue, J.X.; Xu, W.; Zhou, J.C.; Mao, W.G.; Wu, S.S. Effects of High-Temperature Heat Treatment Modification by Impregnation o Physical And Mechanical Properties of Poplar. *Materials* **2022**, *15*, 7334. [CrossRef]
10. Weng, M.Y.; Zhu, Y.T.; Mao, W.G.; Zhou, J.C.; Xu, W. Nano-Silica/Urea-Formaldehyde Resin-Modified Fast-Growing Lumb Performance Study. *Forests* **2023**, *14*, 1440. [CrossRef]
11. Wu, S.S.; Tao, X.; Xu, W. Thermal Conductivity of Poplar Wood Veneer Impregnated with Graphene/Polyvinyl Alcohol. *Fores* **2021**, *12*, 777. [CrossRef]
12. Liu, Q.Q.; Gao, D.; Xu, W. Effect of paint process on the performance of modified poplar wood antique. *Coatings* **2021**, *11*, 117 [CrossRef]
13. Xu, W.; Chen, P.; Yang, Y.; Wang, X.; Liu, X. Effects of freezing and steam treatments on the permeability of populus tomentos *Materialwiss. Werkst.* **2021**, *52*, 907–915. [CrossRef]
14. Tao, M.X.; Liu, X.; Xu, W. Effect of the vacuum impregnation process on water absorption and nail-holding power of silic sol-modified Chinese Fir. *Forests* **2024**, *15*, 261. [CrossRef]
15. Singh, A.P.; Kim, Y.S.; Chavan, R.R. Relationship of wood cell wall ultrastructure to bacterial degradation of wood. *IAWA J.* **201** *40*, 845–870. [CrossRef]
16. Xu, W.; Fang, X.Y.; Han, J.T.; Wu, Z.H. Effect of Coating Thickness on Sound Absorption Property of Four Wood Specie Commonly Used for Piano Soundboards. *Wood Fiber Sci.* **2020**, *52*, 28–43. [CrossRef]
17. Zhang, H.Q.; Feng, X.H.; Wu, Y.; Wu, Z.H. Factors influencing the properties of UV-cured self-matting film. *Prog. Org. Coat.* **202** *189*, 108241. [CrossRef]
18. Zhang, H.Q.; Feng, X.H.; Wu, Y.; Wu, Z.H. Self-matting waterborne polyurethane acrylate wood coating by 222 nm far-UV irradiation in ambient air. *Prog. Org. Coat.* **2024**, *189*, 108305. [CrossRef]
19. Liu, Q.Q.; Gao, D.; Xu, W. Effect of sanding processes on the surface properties of modified poplar coated by primer compare with mahogany. *Coatings* **2020**, *10*, 856. [CrossRef]
20. Liu, Q.Q.; Gao, D.; Xu, W. Effect of polyurethane non-transparent coating process on paint film performance applied on modifie poplar. *Coatings* **2022**, *12*, 39. [CrossRef]
21. Jiang, G.F.; Li, X.F.; Che, Y.L.; Lv, Y.; Liu, F.; Wang, Y.Q.; Zhao, C.C.; Wang, X.J. Antibacterial and anticorrosive propertie of CuZnO@RGO waterborne polyurethane coating in circulating cooling water. *Environ. Sci. Pollut. R.* **2019**, *26*, 9027–904 [CrossRef] [PubMed]
22. Wang, L.; Wang, T.; Hao, R.D.; Wang, Y.M. Construction strategy and mechanism of a novel wood preservative with excellen antifungal effects. *Molecules* **2024**, *29*, 1313. [CrossRef] [PubMed]
23. Vainio-Kaila, T.; Harju, A.; Rohumaa, A.; Paajanen, O.; Venäläinen, M.; Seppä, J.; Veijalainen, A.M.; Pasanen, P. The effect o surface treatment on the antibacterial properties of wood and the possibility to detect the antibacteriality with fluorescenc method. *Forests* **2023**, *14*, 23. [CrossRef]
24. Chang, Y.J.; Wu, Z.H. Synthesized high performance UV-cured wood wax oil using Irgacure 2959 modified thistle oil and linseec oil. *Ind. Crop. Prod.* **2024**, *218*, 118952. [CrossRef]
25. Veiko, A.G.; Olchowik-Grabarek, E.; Sekowski, S.; Roszkowska, A.; Lapshina, E.A.; Dobrzynska, I.; Zamaraeva, M.; Zavodnik, I.B Antimicrobial activity of quercetin, naringenin and catechin: Flavonoids inhibit staphylococcus aureus-induced hemolysis anc modify membranes of bacteria and erythrocytes. *Molecules* **2023**, *28*, 1252. [CrossRef] [PubMed]
26. Wang, L.; Li, W.J.; Li, X.Y.; Liu, J.C.; Chen, Y. Antimicrobial activity and mechanisms of walnut green husk extract. *Molecules* **2023** *28*, 7981. [CrossRef]
27. Tran, N.D.N.; Bui, T.H.; Nguyen, A.P.; Nguyen, T.T.; Nguyen, V.M.; Duong, N.L.; Nguyen, T. The ability of silver-biocha green-synthesized from Citrus maxima peel to adsorb pollutant organic compounds and antibacterial activity. *Green Chem. Lett Rev.* **2022**, *15*, 16–25. [CrossRef]

Irkni, R.; Esmer, O.K. Novel food packaging systems with natural antimicrobial agents. *J. Food Sci. Tech. Mys.* **2015**, *52*, 6095–6111. [CrossRef]

Van, C.K.; Nguyen, P.T.N.; Nguyen, T.T.T.; Bach, L.G. Microencapsulation of Citrus latifolia peel essential oil by spray-drying using maltodextrin: Characterization, antimicrobial activities, and release profile. *Lwt-Food Sci. Technol.* **2024**, *197*, 115825. [CrossRef]

Zhang, Q.Y.; Yang, A.; Tan, W.H.; Yang, W.C. Development, Physicochemical Properties, and Antibacterial Activity of Propolis Microcapsules. *Foods* **2023**, *12*, 3191. [CrossRef]

Huang, N.; Yan, X.X. Preparation of aloe-emodin microcapsules and its effect on antibacterial and optical properties of water-based coating. *Polymers* **2023**, *15*, 1728. [CrossRef] [PubMed]

Zhang, Z.; Zhang, S.S.; Su, R.R.; Xiong, D.; Feng, W.; Chen, J.P. Controlled release mechanism and antibacterial effect of layer-by-layer self-assembly thyme oil microcapsule. *J. Food Sci.* **2019**, *84*, 1427–1438. [CrossRef] [PubMed]

Shilpa, V.S.; Shams, R.; Dash, K.K.; Pandey, V.K.; Dar, A.H.; Mukarram, S.A.; Harsányi, E.; Kovács, B. Phytochemical properties, extraction, and pharmacological benefits of naringin: A review. *Molecules* **2023**, *28*, 5623. [CrossRef] [PubMed]

Chen, S.; Wang, X.J.; Cheng, Y.; Gao, H.S.; Chen, X.H. A review of classification, biosynthesis, biological activities and potential applications of flavonoids. *Molecules* **2023**, *28*, 4982. [CrossRef] [PubMed]

Shangguan, Y.C.; Ni, J.; Jiang, L.L.; Hu, Y.; He, C.B.; Ma, Y.; Wu, G.H.; Xiong, H.J. Response surface methodology-optimized extraction of flavonoids from pomelo peels and isolation of naringin with antioxidant activities by Sephadex LH20 gel chromatography. *Cur. Res. Food Sci.* **2023**, *7*, 100610. [CrossRef] [PubMed]

Lopes, L.A.A.; Rodrigues, J.B.D.; Magnani, M.; de Souza, E.L.; de Siqueira, J.P. Inhibitory effects of flavonoids on biofilm formation by Staphylococcus aureus that overexpresses efflux protein genes. *Microb. Pathog.* **2017**, *107*, 193–197. [CrossRef] [PubMed]

Liu, Z.Y.; Pan, Y.R.; Li, X.S.; Jie, J.X.; Zeng, M.Y. Chemical composition, antimicrobial and anti-quorum sensing activities of pummelo peel flavonoid extract. *Ind. Crop. Prod.* **2017**, *109*, 862–868. [CrossRef]

Valle, J.A.B.; Valle, R.D.S.C.; Bierhalz, A.C.K.; Bezerra, F.M.; Hernandez, A.L.; Arias, M.J.L. Chitosan microcapsules: Methods of the production and use in the textile finishing. *J. Appl. Polym. Sci.* **2021**, *138*, e50482. [CrossRef]

Zhou, J.C.; Xu, W. An aesthetic transparent wood resistant to *Escherichia coli* based on interface optimization. *Eur. J. Wood Wood Prod.* **2023**, *81*, 1569–1579. [CrossRef]

Chen, K.L.; Ni, Y.Z.; Shi, X.; Jia, Z.H.; Qiu, H.; Portale, G. Green fabrication of chitosan microcapsules via double emulsion-simple coacervation and their application in fabrics. *Cellulose* **2023**, *30*, 11861–11873. [CrossRef]

Utami, P.D.; Setianingsih, H.; Tirto Sari, D.R. Microencapsulation, Physicochemical Characterization, and Antioxidant, Antibacterial, and Antiplasmodial Activities of Holothuria atra Microcapsule. *Scientifica* **2024**, *2024*, 5559133. [CrossRef] [PubMed]

Comini, S.; Mandras, N.; Iannantuoni, M.R.; Menotti, F.; Musumeci, A.G.; Piersigilli, G.; Allizond, V.; Banche, G.; Cuffini, A.M. Positive and Negative Ions Potently Inhibit the Viability of Airborne Gram-Positive and Gram-Negative Bacteria. *Microbiol. Spectr.* **2021**, *9*, e00651. [CrossRef] [PubMed]

Monteiro, A.; Cardoso, J.; Guerra, N.; Ribeiro, E.; Viegas, C.; Verde, S.C.; Sousa-Uva, A. Exposure and Health Effects of Bacteria in Healthcare Units: An Overview. *Appl. Sci.-Basel* **2022**, *12*, 1958. [CrossRef]

Tehrani, E.; Amiri, S. Synthesis and characterization PVA electro-spun nanofibers containing encapsulated vitamin C in chitosan microspheres. *J. Text. I.* **2022**, *113*, 212–223. [CrossRef]

Deng, J.Z.; Ding, T.T.; Yan, X.X. Effect of Two Pomelo Peel Flavonoid Microcapsules on the Performance of Waterborne Coatings on the Surface of Poplar Boards. *Coatings* **2024**, *14*, 937. [CrossRef]

GB/T 21866-2008; Test Method and Effect for Antibacterial Capability of Paints Film. Standardization Administration of the People's Republic of China: Beijing, China, 2008.

GB/T 4789.2-2022; National Food Safety Standard Food Microbiological Examination: Aerobic Plate Count. Standardization Administration of the People's Republic of China: Beijing, China, 2022.

GB/T 11186.3-1989; Methods for Measuring the Colour of Paint Films. Part III: Calculation of Colour Differences. Standardization Administration of the People's Republic of China: Beijing, China, 1990.

GB/T 4893.6-2013; Test of Surface Coatings of Furniture. Standardization Administration of the People's Republic of China: Beijing, China, 2013.

Zhang, H.; Zhou, L.M.; Shehzad, H.; Farooqi, Z.H.; Sharif, A.; Ahmed, E.; Habiba, U.; Qaisar, F.; Fatima, N.-E.; Begum, R.; et al. Innovative free radical induced synthesis of WO_3-doped diethyl malonate grafted chitosan encapsulated with phosphorylated alginate matrix for UO_2^{2+} adsorption: Parameters optimisation through response surface methodology. *Sep. Purif. Technol.* **2024**, *353*, 128455. [CrossRef]

Ding, T.T.; Yan, X.X. Preparation process optimization for melamine resin-covered pomelo peel flavonoid antibacterial microcapsules and their effect on waterborne paint film performance. *Coatings* **2024**, *14*, 654. [CrossRef]

coatings

MDPI

Article

Effects of Different Polyols with Functions on the Properties of Polyester Polyol-Based Polyurethane Coatings

Zhaoyang Zhang, Nannan Ni and Yahong Xu *

Key Laboratory for Light-Weight Materials, Nanjing Tech University, Nanjing 210009, China; zzy649509874@163.com (Z.Z.); nan1986@njtech.edu.cn (N.N.)
* Correspondence: 13366615596@163.com

Abstract: Polyester-based polyurethane coatings were widely used in automotive, industrial, construction, and plastics industries due to their excellent mechanical properties, adhesion, and relatively outstanding oil and chemical resistance. In these coatings, the type and ratio of polyester and isocyanate curing agents influenced the cohesion energy, hydrogen bonding, crystallinity, crosslinking density, molecular weight, and morphology of the polyurethane at the microscopic level, thereby affecting the macroscopic mechanical properties, electrical performance, and environmental resistance of the material. However, there was limited systematic research on the effect of crosslinking density on the properties of polyester-based polyurethanes. In this study, an HTP-1 system was composed of neopentyl glycol (NPG) and phthalic anhydride (PA), and an HTP-2 system was composed of neopentyl glycol (NPG), hexahydrophthalic anhydride (HHPA), and adipic acid (AA). A series of polyesters (HTPs) were synthesized by adding polyols with different functional groups and adjusting their proportions in the system. The synthesized polyester was characterized using FT-IR, GPC, and DSC, and then cured with polyisocyanate curing agent N3390 to prepare the coating. The following properties of the films were evaluated: adhesion, impact resistance, pencil hardness, gloss, flexibility, oil resistance, and weather resistance. The results showed that in the HTP-1 system, the introduction of dipentaerythritol resulted in a polyester with a broad molecular weight distribution at high hydroxyl values, with a maximum PDI of 12.66 and a glass transition temperature (Tg) reaching 40.19 °C. The polyesters prepared by introducing three types of multifunctional polyols into the HTP-1 system exhibited good impact resistance, adhesion, and hardness. At low hydroxyl values, the coatings demonstrated good flexibility, but due to the lower crosslinking density, the oil resistance was poor. As the hydroxyl value increased, flexibility decreased, while oil resistance improved. In the HTP-2 system, coatings prepared with three different multifunctional polyols showed good impact resistance, flexibility, and hardness at low hydroxyl values but poor adhesion and oil resistance. As the hydroxyl value increased, adhesion improved from grade 1 to grade 0, and oil resistance improved for coatings prepared with trimethylolpropane and ditrimethylolpropane. However, the oil resistance of coatings prepared with dipentaerythritol decreased. Regarding weather resistance, the HTP-1-series resins primarily exhibited the cleavage of -CH_2 groups, while the HTP-2-series resins showed the cleavage of C-N bonds. Overall, the HTP-2 series resins demonstrated better weather resistance. In the high-hydroxyl-value HTP-2 system, the incorporation of trimethylolpropane or ditrimethylolpropane has been shown to produce coatings that achieve a balance among mechanical properties, flexibility, and oil resistance. This finding provides valuable insights for the design and development of high-performance polyester-based polyurethane coatings.

 https://doi.org/10.3390/coatings150100

Keywords: polyester polyol; resins; polyester

1. Introduction

Polyurethane is one of the most common, versatile, and extensively researched materials worldwide [1,2]. Polyurethane coatings, leveraging their exceptional performance characteristics, have demonstrated significant application value across various fields. Firstly, polyurethane coatings, due to their excellent water resistance, have become a core material in the field of wood protection [3–5], significantly extending the service life of wood. Secondly, their superior weather resistance, aesthetic qualities, and wear resistance have enabled widespread applications in the automotive and construction industries [6–8]. Furthermore, polyurethane coatings, with their outstanding durability, chemical corrosion resistance, and flexibility, provide irreplaceable technical advantages in the field of corrosion protection [9,10], ensuring reliable and long-lasting protection for equipment and infrastructure under complex and demanding environmental conditions.

Based on their chemical structure, polyurethane coatings can be classified into polyester-based polyurethane coatings and polyether-based polyurethane coatings [11,12]. Compared to polyether-based coatings, polyester-based coatings exhibit higher glass transition temperatures and superior properties. Due to the flexibility of ether bonds, polyether-based polyurethane films demonstrate better cold resistance and alkaline hydrolysis resistance, as well as lower viscosity. However, polyether-based polyurethanes lag behind polyester-based ones in terms of oil resistance, water resistance, mechanical properties, drying performance, and reactivity with -NCO. Consequently, polyester-based polyurethanes find far more extensive application in the coatings industry.

Polyester-based polyurethanes are synthesized through the reaction of polyester polyols with polyisocyanates, with the properties of the resulting polyurethane being predominantly determined by the chemical composition and structure of the polyester polyols. When the type of polyisocyanate is fixed, the characteristics of the polyester polyol dictate the final performance of the coating.

During polyester synthesis, the selection of polyols is critical. Di-functional alcohols increase the molecular weight of the resin, while polyols with functionalities above three increase the hydroxyl value and branching degree of the polyester resin. Branching in polyester resins relies primarily on hydroxyl groups from tri-functional alcohols. To achieve faster drying, trimethylolpropane is commonly used during polyester synthesis, whereas glycerol, containing secondary hydroxyl groups, is less frequently employed. A higher proportion of diols and fewer triols, along with minimal free hydroxyl groups, yield coatings with excellent flexibility but relatively lower hardness and chemical resistance. Conversely, using a higher proportion of triols increases free hydroxyl groups, leading to coatings with higher crosslinking density, enhanced hardness, and better chemical resistance, albeit with higher viscosity and lower solvent resistance.

In some studies, the use of tetra-functional polyols such as ditrimethylolpropane to replace traditional tri-functional polyols such as trimethylolpropane has been explored to prepare coatings with superior properties. In alkyd resin systems, hexafunctional polyols such as dipentaerythritol are often used as raw materials. However, research on the use of highly functional polyols such as dipentaerythritol for polyester synthesis remains relatively limited, presenting significant potential for further exploration.

Jiang [13] synthesized ditrimethylolpropane acrylate using ditrimethylolpropane and acrylic acid as raw materials. Their experimental results demonstrated that ditrimethylolpropane acrylate exhibited faster curing rates, higher hardness, and better flexibility compared to trimethylolpropane acrylate.

Huang et al. [14] utilized a "one-pot" vacuum polymerization method to synthesize hydroxyl-terminated polyesters for two-component polyurethane coatings using ditrimethylolpropane, DMPA, monocarboxylic acids, and caprolactone. These polyesters, with solid concentrations of 100%, showed viscosities (3304–9060 mPa·s) lower than those of linear polymers with equivalent molecular weights and significantly reduced intrinsic viscosities. Films cured from these hydroxyl polyesters exhibited excellent impact resistance, high adhesion strength, high hardness, and minimal weight loss of only 5% at 280 °C.

Haseebuddin et al. [15] prepared high-solid alkyd resins by esterifying soybean fatty acids (SOFAs) and dehydrated castor oil fatty acids (DCOFAs) with dipentaerythritol (DPE). Their analysis of the physical, optical, thermal, and mechanical properties revealed that increasing DPE content improved surface distribution, formed more dendritic network structures, enhanced flowability, and increased coating hardness.

Baloji Naik [16] used DPE as the primary raw material and 2,2-bis(hydroxymethyl)propionic acid (BMPA) as a chain extender to synthesize second-generation hyperbranched polyols (HBPs). These were further reacted with various concentrations of linseed oil fatty acids (LOFAs) to produce a series of hyperbranched alkyd (HBA) resins. These resins exhibited excellent adhesion, high gloss, and remarkable flexibility.

In this study, neopentyl glycol and phthalic anhydride were used as the HTP-1 system, and neopentyl glycol, hexahydrophthalic anhydride, and adipic acid were used as the HTP-2 system. A series of polyesters (HTPs) were synthesized by adding polyols with different functional groups and adjusting their proportions in these systems. The polyester was characterized using FT-IR, GPC, and DSC, and cured with polyisocyanate curing agent N3390 to prepare the film. The adhesion, impact resistance, pencil hardness, gloss, flexibility, oil resistance, and weather resistance of the film were evaluated. This study investigates the impact of different functionalities of polyols on the properties of polyester polyurethane coatings, aiming to explore the effect of crosslinking density on the performance of polyester-based polyurethane coatings, thus achieving a balance between mechanical properties and resistance.

2. Experiments

2.1. Materials

Neopentyl glycol (NPG, industrial grade) was supplied by Jilin Chemical BASF (Jilin, China).; trimethylolpropane (TMP, industrial grade) was obtained from Jilin Petrochemical (Changchun, China); phthalic anhydride (PA, industrial grade) was purchased from Nanjing Chemical Reagent (Nanjing, China); ditrimethylolpropane (Di-TMP, industrial grade) and dipentaerythritol (Di-PE, industrial grade) were provided by Chifeng Ruiyang Chemical (Chifeng, China); hexahydrophthalic anhydride (HHPA, industrial grade) and adipic acid (AA, industrial grade) were supplied by Suzhou Chuanyuan Chemical (Suzhou, China); monobutyltin oxide (FC4100, industrial grade) was obtained from Beijing Zhengheng (Beijing, China); and n-butyl acetate (BAC), cyclohexanone (CYC), and xylene (industrial grade) were sourced from Zibo Kangping Economic and Trade (Zibo, China) Hexamethylene diisocyanate (HDI) trimer (Desmodur N3390, with an NCO content of 19.36%) was provided by Bayer AG (Leverkusen, Germany). The leveling agent used was BYK311 (Leverkusen, Germany).

2.2. Synthesis of Hydroxyl-Terminated Polyesters

We added the formula-specified amounts of polyacid, polyol, and other materials to a three-neck flask equipped with a water separator, reflux condenser, agitator, and thermometer. A small amount of mono-butyl tin oxide and reflux xylene was then added. The mixture was heated to 70–90 °C to melt most of the components. Subsequently, the mixture was stirred and gradually heated to 140–160 °C to initiate reflux. Simultaneously, the column top temperature was maintained between 95 and 105 °C. The temperature was increased gradually by 6–8 °C per hour, reaching 220 °C. At 220 °C, samples were taken to test the acid value, ensuring it was below 10 mg KOH/g. Nitrogen was then introduced into the system, and after 30 min, the mixture was cooled to approximately 150 °C. Finally, a mixture of cyclohexanone and butyl acetate was added for dilution, and the material was discharged while still hot.

The formula of polyester synthesis has theoretical hydroxyl value mass percentage as shown in Table 1. The polyester synthesis diagram is shown in Scheme 1.

Table 1. Ratios of the raw materials used to synthesize HTPs and the corresponding theoretical hydroxyl mass ratios.

HTPs	Molar Ratios of Raw Materials	Theoretical Hydroxyl Mass Ratio
HTP-1-0	NPG:PA = 1.15:1	2
HTP-1-1	NPG:TMP:PA = 1.05:0.17:1	4
HTP-1-2	NPG:Di-TMP:PA = 1.12:0.1:1	4
HTP-1-3	NPG:Di-PE:PA = 1.06:0.08:1	4
HTP-1-4	NPG:TMP:PA = 0.12:1.03:1	8
HTP-1-5	NPG:Di-TMP:PA = 0.34:0.75:1	8
HTP-1-6	NPG:Di-PE:PA = 0.87:0.27:1	8
HTP-2-0	NPG:HHPA:AA = 2.29:1:1	2
HTP-2-1	NPG:TMP:HHPA:AA = 2.21:0.28:1:1	4
HTP-2-3	NPG:Di-TMP:HHPA:AA = 2.44:0.09:1:1	4
HTP-2-5	NPG:Di-PE:HHPA:AA = 2.17:0.15:1:1	4
HTP-2-2	NPG:TMP:HHPA:AA = 0.29:2:1:1	8
HTP-2-4	NPG:Di-TMP:HHPA:AA = 1.26:1.16:1:1	8
HTP-2-6	NPG:Di-PE:HHPA:AA = 2.43:0.31:1:1	8

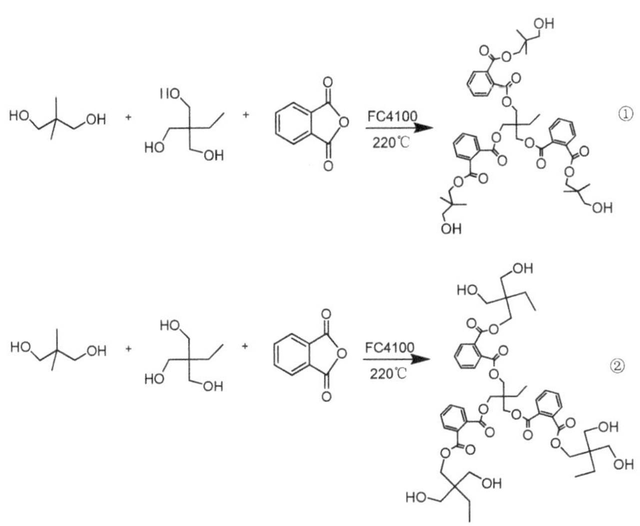

Scheme 1. Diagram of resin synthesis route for HTP-1-1 and HTP-1-4.

① Schematic diagram of the synthesis of HTP-1-1, which is a polyester system with a low hydroxyl value; ② schematic diagram of the synthesis of HTP-1-4, which is a polyester system with a high hydroxyl value. In the HTP-1-1 resin, the amount of neopentyl glycol (NPG) is higher, while the trimethylolpropane (TMP) content is lower. When the three hydroxyl groups on TMP react with phthalic anhydride (PA), the carboxyl groups formed subsequently react with the hydroxyl groups from the abundant NPG. Conversely, in the HTP-1-4 resin, the amount of TMP is higher, and NPG content is reduced. The three hydroxyl groups on trimethylolpropane (TMP) react with phthalic anhydride (PA) to form carboxyl groups, which subsequently undergo reactions with the hydroxyl groups present on additional TMP molecules. This is the key difference between the two reactions. The composite diagram of HTP-1-2 and HTP-1-1 is structurally similar, as is the composite diagram of HTP-1-5 and HTP-1-4. In both cases, the TMP nucleus is replaced by Di-TMP. Similarly, the synthesis of HTP-1-3 and HTP-1-6 resembles that of HTP-1-1, except the TMP nucleus is replaced by Di-PE. The same pattern applies to the HTP-2 systems.

2.3. *Preparation of Polymeric Film*

Hydroxyl polyester was prepared in a mixture of cyclohexanone (CYC) and butyl acetate (BAC) (CYC:BAC = 2). Then, N3390 was mixed in a molar ratio of NCO:OH = 1.1:1 in the presence of the catalyst dibutyltin dilaurate (DBTDL) (0.01 wt % reactant and solvent) and the leveling agent (0.1 wt % reactant and solvent) to maintain a solid content of 40–41 wt %. Then, we sprayed on an aluminum plate followed by a tinplate using a spray gun to prepare the coating film, and we tested the coating performance after 7 days of normal-temperature curing.

2.4. *Characterization*

2.4.1. Fourier Transform Infrared Spectroscopy (FT-IR)

The hydroxyl resin sample was coated on the pressed KBr chip using the KBr pressing method, and then dried under an infrared lamp. The infrared spectrum of the sample was collected using an FTIR-8100 Fourier-transform infrared spectrometer from Shimadzu Company, (Kyoto, Japan). The scanning range was 4000~400 cm^{-1}, and the scanning was performed 32 times. The resolution was 2 cm^{-1}.

2.4.2. Gel Permeation Chromatography (GPC)

For the gel chromatography (Breez2 type) analysis, the separation column used was a PLGEL 10MIXEDx3 (Waters, Milford, MA, USA) high-crosslinked spherical polystyro-divinylbenzene polymer column. The mobile phase was 4 (THF) and the flow rate was 1.500 mL/min; the standard material was monodisperse polyethylene; and the detector was an HP1047A differential refractometer (Waters, Milford, MA, USA).

2.4.3. Differential Scanning Calorimetry (DSC)

Differential scanning calorimetry (DSC) was measured with an American UniversalV1.717 TA Instruments DSC 2910 thermal analyzer (TA, New Castle, DE, USA), and the heating rate was 10 °C/min. The test temperature range was −80 °C–80 °C, the ambient atmosphere was nitrogen, and the sample size was 5–10 mg.

2.4.4. Heat Resistance of Hydroxyl Resin

All polyester samples were then dried at 180 °C for a sufficient period to thoroughly remove any solvents. The mass of each polyester sample after drying at 180 °C was taken as the baseline. Subsequently, the samples were further dried at each of the following temperatures for 3 h: 220 °C, 260 °C, and 300 °C.

The mass retention rate is denoted as u. The mass retention rate at 220 °C, denoted as $u_{200°C}$, is calculated as follows:

$$u_{220°C} = \frac{m_{220°C}}{m_{180°C}} \times 100$$

where $m_{220°C}$ and $m_{180°C}$ are the masses of the polyester samples after drying at 200 °C and 180 °C, respectively.

2.4.5. Properties of the Polymeric Film

The hardness of the film was measured according to GB/T 6739-2006, and adhesion was tested according to GB/T 9286-1998, flexibility according to GB/T 1731-1993, and impact resistance according to GB/T 1732-1993. Polyester viscosity was tested according to GB/T 1723-1993 and polyester solids were tested according to GB/T 1725-2007.

The polyester varnish sprayed onto the aluminum plate was placed in No. 15 aviation hydraulic oil and soaked for 7 days. After 7 days, we observed the change in the shape and hardness of the varnish surface on the aluminum plate.

The UVB-313 aging chamber(Dongguan Haotian test equipment, Dongguan, China) was utilized to simulate accelerated ultraviolet (UV) radiation-induced degradation in accordance with the GB/T 14522-2008 standard. The experimental conditions were meticulously controlled, with a temperature of 60 °C, a radiant irradiance of 0.71 W/m^2, and a relative humidity (RH) maintained at ≥95%. Hot air convection was employed to ensure uniform exposure during continuous irradiation. The total duration of exposure was 400 h, with sample assessments conducted at 100 h intervals. To evaluate photodegradation, the color difference ΔE between the irradiated and unirradiated regions of the sample was quantitatively analyzed using a BYK 7075 color difference meter (BYK, Wesel, Germany), and GB/T250-1989 was referred to for the measurement of color difference. In parallel, Fourier-transform infrared (FT-IR) spectroscopy was performed using an FTIR8100 spectrometer (Shimadzu, Kyoto, Japan) to investigate the alterations in the molecular structure of the sample surfaces. Shifts in characteristic absorption bands were monitored to elucidate the chemical modifications induced by prolonged UV exposure, providing insight into the underlying mechanisms of surface degradation over time.

3. Results and Discussion

3.1. Structural Characterization of HTPs

The infrared spectrogram of the HTP-1 system is shown in Figure 1, where the widest and strongest -OH peaks are observed near 3520 cm^{-1} because of the residual active hydroxyl group in the polyol. In the polyester systems containing phthalic anhydride, the -CH=CH$_2$ stretching vibration of the benzene ring occurs at 3060 cm^{-1}. The peaks at 2855 and 2927 cm^{-1} correspond to aliphatic C-H stretching vibrations, and the characteristic peaks of esters can be seen in the spectrum at 1720 cm^{-1}. The C=C stretching vibration peak of the benzene ring group is observed at 1600 cm^{-1}. The peaks at 1375 and 1460 cm^{-1} correspond to symmetric and asymmetric bending vibrations of methyl groups. The peak at 1300–1250 cm^{-1} is the stretching vibration of C-O-C, the peak at 980 cm^{-1} and 864 cm^{-1} is the bending vibration peak of C-O-C, and the =CH out-of-plane bending peak at 708 cm^{-1} and 737 cm^{-1} is the benzene ring.

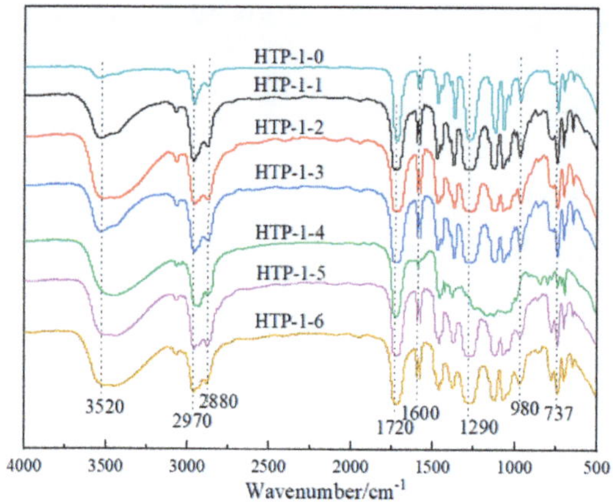

Figure 1. Infrared spectrum of the HTP-1 system.

The infrared spectrum of the HTP-2 system is shown in Figure 2. The widest and strongest -OH peak is also observed near 3530 cm^{-1}, the peaks at 2960 and 2870 cm^{-1} correspond to the aliphatic C-H stretching vibration peak, and the characteristic peak of ester is also seen in the spectrum at 1720 cm^{-1}. The infrared spectra of the two systems prove that the polyester reaction occurred, and with an increase in the hydroxyl value of the reaction product, the intensity of the hydroxyl value peak increases gradually at 3520 cm^{-1}.

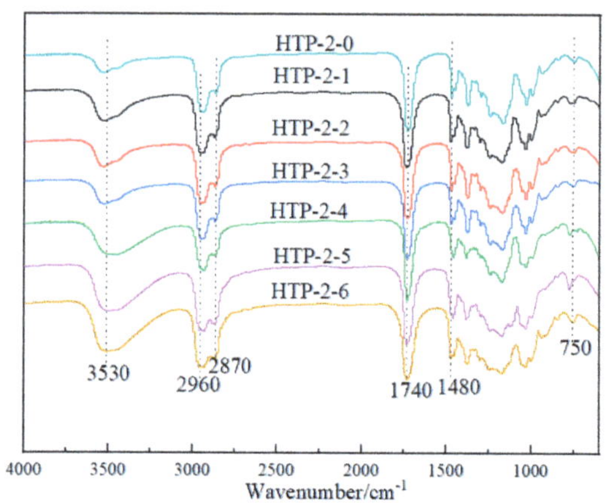

Figure 2. Infrared spectrum of the HTP-2 system.

3.2. Molecular Weight Distribution of HTPs

The molecular weight data for the HTP-1 and HTP-2 series of polyesters are presented in Table 2, with their molecular weight distributions illustrated in Figures 3 and 4, respectively. Both systems show that, despite the relatively similar molecular weights among certain samples (e.g., HTP-1-2 vs. HTP-1-3, and HTP-2-1 vs. HTP-2-3), notable differences in PDI are observed, which can be attributed to steric hindrance effects. Specifically, the bulky groups in dipentaerythritol and the resulting steric hindrance between adjacent groups reduce the reactivity of hydroxyl groups and constrain chain segment mobility, leading to broader molecular weight distributions. In the HTP-1 series, the rigid aromatic ring structure further amplifies this effect by introducing additional spatial hindrance.

Table 2. Molecular weight distribution of the HTP system.

HTPs	Mn	Mw	Mp	Mz	Mz/Mw	PDI
HTP-1-0	1910	3624	3652	5822	1.6	1.89
HTP-1-1	1846	3457	2811	5720	1.65	1.87
HTP-1-2	2932	14,017	5946	42,533	3.03	4.78
HTP-1-3	3209	31,648	5493	132,151	4.18	9.86
HTP-1-4	2267	6033	4825	12,689	2.1	2.66
HTP-1-5	3168	15,946	6557	50,053	3.14	5.03
HTP-1-6	3415	43,249	5066	178,604	4.13	12.66
HTP-2-0	2441	4779	4848	7563	1.58	1.95
HTP-2-1	2139	4378	3562	7609	1.74	2.05
HTP-2-2	2069	4292	3357	7692	1.79	2.08
HTP-2-3	2198	7146	2614	17,842	2.5	3.25
HTP-2-4	3450	30,832	7452	127,856	4.15	8.94
HTP-2-5	3405	24,970	6825	97,844	3.92	7.33
HTP-2-6	2881	13,235	3713	42,266	3.19	4.59

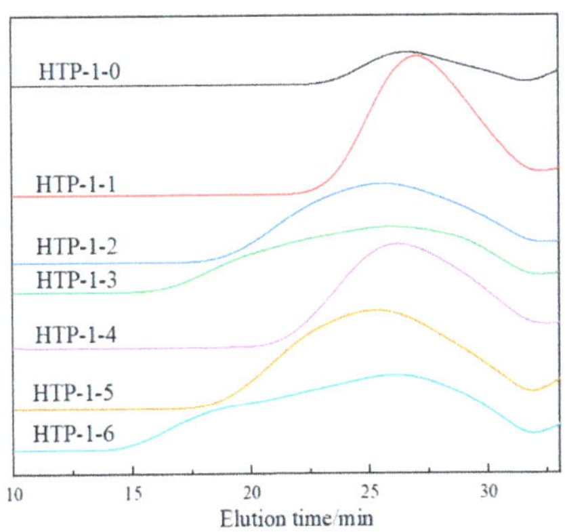

Figure 3. Molecular weight distribution curve of the HTP-1 system.

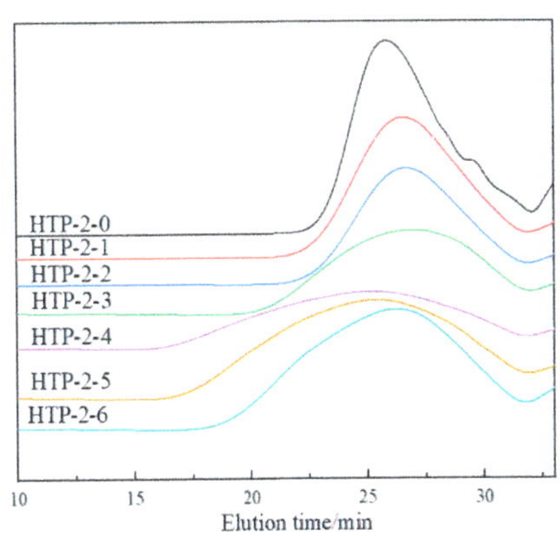

Figure 4. Molecular weight distribution curve of the HTP-2 system.

As the hydroxyl value increases, the molecular weight distributions in both systems become progressively broader. For instance, in the HTP-1 series, the PDI of HTP-1-6 rises significantly to 12.66. Similarly, in the HTP-2 series, the PDI of HTP-2-4 increases from 2.05 to 8.94, reflecting the introduction of more reactive centers that generate both low-molecular-weight segments and greater molecular weight heterogeneity.

3.3. Analysis of Glass Transition Temperature of HTPs

The differential scanning calorimetry (DSC) analysis results of the HTP-1 resin system are presented in Figure 5 and Table 3. Upon introducing three types of multifunctional polyols, the glass transition temperature (Tg) exhibits an overall increasing trend. Due to the inherent symmetry of neopentyl glycol, the HTP-1-0 linear polyester prepared with this polyol crystallizes at room temperature, which affects its Tg. Under identical hydroxyl value conditions, the polyester synthesized with dipentaerythritol achieves the highest Tg, reaching 40.19 °C, which can be attributed to the complex structure of dipentaerythritol and the rigidity of phthalic anhydride. Furthermore, as the hydroxyl value increases, the Tg of polyesters prepared with all three multifunctional polyols also increases.

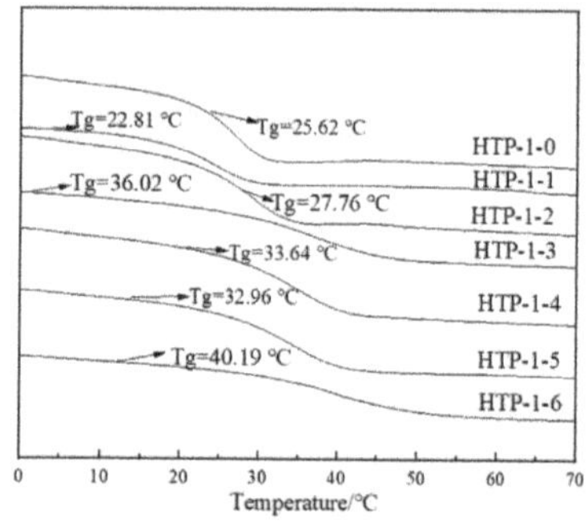

Figure 5. Glass transition temperature of the HTP-1 system.

Table 3. Glass transition temperature of the HTP system.

HTP-1	Tg	HTP-2	Tg
HTP-1-0	25.62 °C	HTP-2-0	−23.04 °C
HTP-1-1	22.81 °C	HTP-2-1	−18.34 °C
HTP-1-2	27.76 °C	HTP-2-2	−21.23 °C
HTP-1-3	36.02 °C	HTP-2-3	−19.40 °C
HTP-1-4	33.64 °C	HTP-2-4	−11.32 °C
HTP-1-5	32.96 °C	HTP-2-5	−10.52 °C
HTP-1-6	40.19 °C	HTP-2-6	−18.35 °C

The DSC analysis results for the HTP-2 resin system are shown in Figure 6 and Table 3. This system lacks rigid monomers such as aromatic rings and incorporates a flexible monomer—adipic acid—resulting in polyesters with Tg values below 0 °C. Among them, the Tg of HTP-2-0 polyester with a hydroxyl value of 2 is −23.04 °C. Introducing three different multifunctional polyols leads to an increase in hydroxyl value and a corresponding rise in Tg. Under identical hydroxyl value conditions, dipentaerythritol, due to its higher hydroxyl content and relatively low usage amount, produces polyesters with a lower Tg compared to the other two polyols. Even with increased hydroxyl values, the Tg of dipentaerythritol-based polyesters exhibits only a modest rise. In contrast, the polyester prepared with ditrimethylolpropane demonstrates the highest Tg, reaching −10.52 °C, which can be attributed to the symmetrical and complex distribution of ditrimethylolpropane's molecular structure. At the same hydroxyl value, the amount of ditrimethylolpropane used is slightly lower than that of trimethylolpropane, and the higher hydroxyl group content in its structure contributes to more complex reaction distributions, which is the primary factor driving the Tg enhancement.

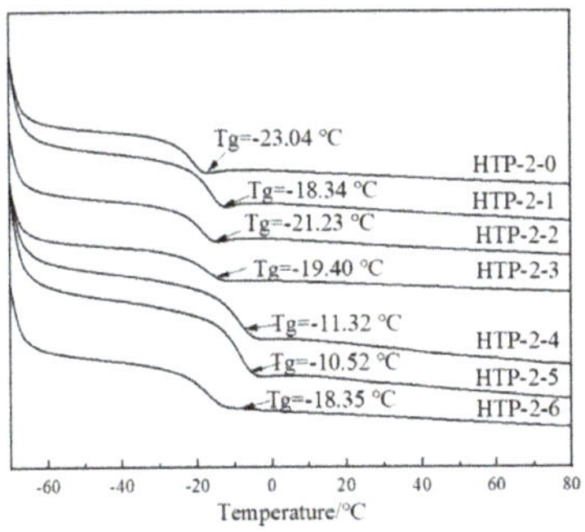

Figure 6. Glass transition temperature of the HTP-2 system.

3.4. *Analysis of Heat Resistance of HTPs*

The quality retention rates of HTP-1 and HTP-2 series resins at different temperatures are shown in Table 4, while the time-dependent quality retention rates of the HTP-1 series are illustrated in Figure 7. The temperature-dependent quality retention rates of the HTP-2 polyester series are presented in Figure 8. At 300 °C, the quality retention rate of the linear resin HTP-1-0 is 86.96%. Under conditions of low hydroxyl value, the introduction of three polyols with different degrees of functionality leads to an improvement in quality retention. Among these, HTP-1-3 achieves the highest quality retention rate of 90.63%. This enhancement can be attributed to the high crosslinking density generated by the multiple active sites of dipentaerythritol, along with the significant steric hindrance and the large cyclic structure of the benzene ring, which collectively contribute to the improved thermal stability at elevated temperatures.

Table 4. HTP system mass retention ratio.

HTP-1	180 °C	220 °C	260 °C	300 °C	HTP-2	180 °C	220 °C	260 °C	300 °C
HTP-1-0	100%	98.52%	92.96%	86.96%	HTP-2-0	100%	98.13%	95.49%	88.01%
HTP-1-1	100%	99.11%	93.28%	87.16%	HTP-2-1	100%	97.66%	95.54%	91.52%
HTP-1-2	100%	98.89%	93.85%	87.95%	HTP-2-2	100%	97.35%	95.01%	89.95%
HTP-1-3	100%	99.13%	97.15%	90.63%	HTP-2-3	100%	98.08%	96.06%	92.12%
HTP-1-4	100%	99.20%	96.93%	89.19%	HTP-2-4	100%	98.77%	96.50%	92.81%
HTP-1-5	100%	97.87%	93.05%	86.45%	HTP-2-5	100%	97.68%	95.29%	91.15%
HTP-1-6	100%	98.94%	96.69%	92.17%	HTP-2-6	100%	97.51%	95.08%	91.66%

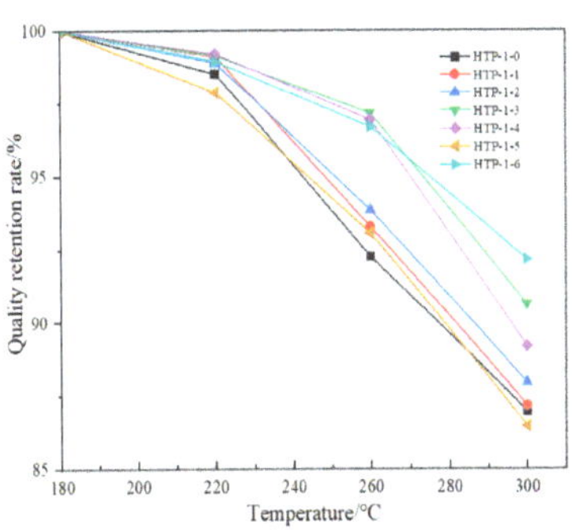

Figure 7. Quality retention rate of the HTP-1 system varies with temperature.

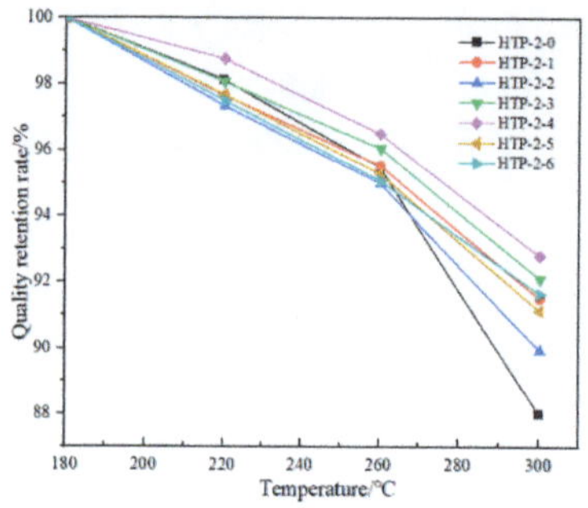

Figure 8. Quality retention rate of the HTP-2 system varies with temperature.

As the hydroxyl value increases, the quality retention rates of HTP-1-4 and HTP-1-6 are further enhanced, with HTP-1-6 reaching a retention rate of 92.17%. This result correlates with its higher glass transition temperature. However, the quality retention rates of HTP-1-2 and HTP-1-5 are lower than those of the other samples, potentially due to the presence of a greater number of ether bonds in their molecular structures. This observation is consistent with findings by Zhang et al. [17,18], who investigated the thermal degradation behavior of glycerol-adipate polyester and trimethylolpropane-adipate polyester using infrared spectroscopy. Their study revealed that during thermal degradation, the strength of the carbon–oxygen bond, as indicated by the infrared absorption peak at 1173 cm^{-1}, gradually increases, suggesting that the degradation process tends to favor the formation of ether bonds. In polyester systems, the existing abundance of ether bonds, which are prone to rotation, accelerates the rate of thermal degradation, thereby leading to lower quality retention rates.

The temperature-dependent quality retention rates of the HTP-2 polyester series are presented in Figure 8. At 300 °C, HTP-2-0 exhibits a quality retention rate of 88.01%. Upon the introduction of three polyols with varying degrees of functionality, the overall quality retention rates exceed 90%, except for HTP-2-2, whose quality retention rate remains below this threshold. As the hydroxyl value increases, the quality retention rate at 300 °C improves correspondingly.

Compared to the HTP-1 system, the HTP-2 system demonstrates slightly higher quality retention rates. This can be attributed to the presence of a greater proportion of flexible chain segments and the absence of bulky steric hindrance, such as that caused by benzene rings. The reduced steric constraints facilitate higher chain segment mobility and reactivity during crosslinking, leading to the formation of a more thermally stable network structure. Consequently, the HTP-2 system exhibits superior thermal resistance characteristics.

3.5. Mechanical Properties of Polymeric Film

The mechanical performance test results of HTP-1 polyester are presented in Table 5. This system demonstrates excellent impact resistance, with all samples achieving an adhesion level of 0 and a pencil hardness of 6H. This high hardness is primarily attributed to the incorporation of phthalic anhydride-containing aromatic rings. The rigid structure of the benzene ring significantly enhances the mechanical strength of the polyester coating, which corresponds to its superior pencil hardness.

Table 5. Mechanical properties of HTP system.

HTPs	Adhesion (Grade)	Flexibility (mm)	Impact Resistance (cm)	Pencil Hardness	Gloss (60°)
HTP-1-1	0	1	50	6H	95
HTP-1-2	0	1	50	6H	96
HTP-1-3	0	1	50	5H	95
HTP-1-4	0	3	50	6H	92
HTP-1-5	0	3	50	6H	97
HTP-1-6	0	1	50	6H	93
HTP-2-1	1	1	50	2H	93
HTP-2-2	1	1	50	2H	94
HTP-2-3	1	1	50	3H	92
HTP-2-4	0	2	50	3H	92
HTP-2-5	0	2	50	4H	91
HTP-2-6	0	1	50	4H	96

Under low-hydroxyl-value conditions, the polyester coatings prepared with the three different multifunctional polyols exhibit a flexibility of 1 mm, indicating excellent flexibility. However, at higher hydroxyl values, the flexibility of the coatings prepared with trimethylolpropane (HTP-1-4) and ditrimethylolpropane (HTP-1-5) decreases to 3 mm, whereas the flexibility of the coatings prepared with dipentaerythritol (HTP-1-6) remains at 1 mm. This phenomenon is closely related to the high functionality and significant steric hindrance of dipentaerythritol. At the same hydroxyl value, the lower usage amount of dipentaerythritol results in predominantly linear polyester chains with lower branching degrees. This reduced branching minimizes excessive crosslinking during the curing process, thereby maintaining a balance between high crosslink density and good flexibility. Consequently, even under high-hydroxyl-value conditions, polyester coatings prepared with dipentaerythritol exhibit superior flexibility.

The mechanical performance test results of the HTP-2 polyester system are summarized in Table 5. All samples exhibit good impact resistance; however, the pencil hardness of the HTP-2 polyesters is lower compared to that of the HTP-1 system. At low hydroxyl values, the adhesion of the coatings prepared with the three multifunctional polyols decreases to grade 1. The adhesion of a coating is generally influenced by factors such as the molecular weight of the base resin, the presence of polar groups, and the degree of crosslinking. Under low-hydroxyl-value conditions, the usage of multifunctional polyols is relatively low, resulting in predominantly linear molecular chains with low branching degrees. Consequently, the polyester lacks sufficient unreacted hydroxyl groups to effectively crosslink with the curing agent, leading to a lower crosslink density and reduced adhesion.

At higher hydroxyl values, the increased incorporation of multifunctional polyols significantly enhances the branching degree of polyester molecules, thereby increasing the crosslink density and improving adhesion to grade 0. While the increased usage of multifunctional polyols may reduce flexibility due to higher crosslink density, the incorporation of flexible adipic acid into the HTP-2 system mitigates this effect. The long-chain structure and flexibility of adipic acid help to counterbalance excessive rigidity within the system, maintaining the coating flexibility at 2 mm. This indicates that the strategic introduction of flexible monomers can simultaneously enhance crosslink density and improve the flexibility of polyester coatings.

3.6. Oil Resistance of Polymeric Film

The oil resistance data of HTP system resin are shown in Table 6. In the HTP-1 system, under low-hydroxyl-value conditions, the polyester coatings prepared using trimethylolpropane (TMP) and ditrimethylolpropane (DTMP) exhibit delamination after oil immersion, with their hardness decreasing from 6H to HB. In contrast, the coatings prepared with dipentaerythritol (DPE) remain intact even under low-hydroxyl-value conditions, with only a slight reduction in hardness. This can be attributed to the synergistic effect of the rigid benzene ring structures and the complex molecular structure of DPE, which hinder oil penetration into the coating and prevent swelling. As the hydroxyl value increases, the reduction in hardness of the coatings prepared from all three polyfunctional polyols after oil immersion decreases significantly, and the surface condition of the coatings remain intact. This is because higher hydroxyl values increase the polyol content, providing more active hydroxyl groups to react with curing agents, thereby enhancing the crosslinking density and effectively preventing oil penetration.

Table 6. HTP system oil resistance results.

HTPs	Initial Pencil Hardness	Pencil Hardness After Oil Immersion	Oil-Soaked State
HTP-1-1	6H	HB	Partial shedding
HTP-1-2	6H	H	Partial shedding
HTP-1-3	5H	3H	Surface integrity
HTP-1-4	6H	6H	Surface integrity
HTP-1-5	6H	5H	Surface integrity
HTP-1-6	6H	6H	Surface integrity
HTP-2-1	2H	/	Completely shedding
HTP-2-2	2H	/	Completely shedding
HTP-2-3	3H	/	Completely shedding
HTP-2-4	3H	3H	Surface integrity
HTP-2-5	4H	3H	Surface integrity
HTP-2-6	4H	/	Completely shedding

In the HTP-2 system, under low-hydroxyl-value conditions, the polyester coatings prepared using all three types of polyfunctional polyols were unable to resist oil penetration, resulting in noticeable swelling. This is primarily due to the absence of rigid benzene ring structures in the HTP-2 system and its inherently lower crosslinking density, which fails to effectively prevent oil ingress. As the hydroxyl value increases, more active hydroxyl groups participate in reactions with the curing agents, leading to a significant increase in crosslinking density and an improvement in the coating surface condition, thereby enhancing oil resistance. However, under equivalent hydroxyl-value conditions, the oil resistance of the coatings prepared with DPE decreases. Without the rigid benzene ring structure, the steric hindrance effect of DPE limits the reactivity of some active hydroxyl groups with curing agents, reducing the achievable crosslinking density and thereby impairing the oil resistance of the coating.

3.7. Weather Resistance (QUV) of Polymeric Film

The color difference data of the HTP series resins after UV aging for different durations are presented in Table 7. When $\Delta E \leq 1$, the naked eye cannot perceive any noticeable difference. When ΔE is between 1 and 2, the naked eye can perceive a slight color difference. As ΔE increases to between 2 and 3.5, the color difference becomes more noticeable. When ΔE exceeds 3.5, the color difference is particularly pronounced. The data indicate that the HTP-1-series resins exhibit significantly larger color differences after UV exposure, with values generally exceeding 10. Moreover, the color difference increases progressively with aging time. In contrast, the HTP-2 series resins display much smaller color differences, with most values ranging from 1 to 3.

Table 7. The color difference (ΔE) in the HTP system changes with time.

HTP-1/min	100 h	200 h	300 h	400 h	HTP-2/min	100 h	200 h	300 h	400 h
HTP-1-0	9.31	9.66	11.71	12.53	HTP-2-0	0.35	0.62	0.94	1.48
HTP-1-1	8.50	8.68	11.51	14.93	HTP-2-1	0.76	1.16	1.53	2.04
HTP-1-2	6.40	10.03	11.20	12.05	HTP-2-2	0.54	1.00	1.19	1.86
HTP-1-3	9.41	11.43	10.33	16.27	HTP-2-3	0.63	0.67	0.82	1.33
HTP-1-4	4.66	7.29	8.86	10.32	HTP-2-4	1.07	1.57	1.90	2.26
HTP-1-5	7.56	9.72	10.90	13.11	HTP-2-5	0.33	0.43	1.11	2.53
HTP-1-6	7.75	8.32	10.16	10.26	HTP-2-6	0.67	1.08	1.26	3.09

For the HTP-1 system, UV irradiation induces chain scission within the polyurethane, and the presence of oxygen accelerates this process. Under UV light, oxygen generates a significant amount of reactive oxygen species, which act on the reactive sites along the polymer chains. This leads to the formation of various chromophoric and auxochromic groups, such as quinones and stilbene quinones, resulting in a pronounced increase in color difference [19,20].

In contrast, the HTP-2 system shows smaller color differences, which can be attributed to differences in molecular structure. The absence of aromatic structures in HTP-2 reduces its susceptibility to photooxidation. Furthermore, the lower reactivity of the alpha hydrogen atoms in the polyol structure effectively minimizes the formation of chromophores, thereby enhancing UV stability and leading to a smaller color difference.

The infrared spectra of the HTP-1-series polyesters after 400 h of UV aging are shown in Figure 9. In the HTP-1-0 system, the peaks at 2940 cm^{-1} and 2850 cm^{-1} correspond to the absorption vibrations of the $-CH_2-$ and $-CH_3$ groups, respectively, while the peak at 1687 cm^{-1} corresponds to the absorption of the carbonyl group. The peaks at 1261 cm^{-1}, 1130 cm^{-1}, and 1078 cm^{-1} correspond to the stretching vibrations of the urethane bond in the curing agent, ester group in the polyester, and ether bond (C-O-C) in the polyester, respectively. In all systems except HTP-1-0, a decrease in the intensity of the $-CH_2-$ and $-CH_3$ absorption peaks at 2940 cm^{-1} and 2850 cm^{-1} is observed, indicating partial scission of the $-CH_2-$ groups. Meanwhile, an increase in the intensity of the absorption peak at 1078 cm^{-1} is observed, which is associated with the stretching vibration of C-O in polyurethane. This indicates that under UV irradiation, some chemical bonds in the polyurethane molecules undergo scission, resulting in the formation of new oxidation products, which leads to an increase in C-O groups [21].

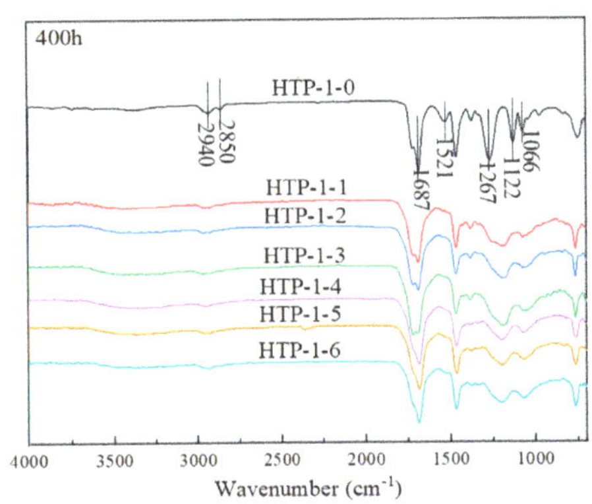

Figure 9. Infrared spectrum of the HTP-1-series resin after UV aging for 400 h.

The infrared spectrum of the HTP-2 system after 400 h of aging is shown in Figure 10. In HTP-2-0, absorption peaks are observed at 2940 cm^{-1} and 2863 cm^{-1}, corresponding to the stretching vibrations of the -CH_2 and -CH_3 groups, respectively. The peak at 1725 cm^{-1} represents the C=O stretching vibration of the polyurethane, while the peak at 1531 cm^{-1} corresponds to the -NH group. The absorption peak at 1251 cm^{-1} is attributed to the amide group, and the peak at 1170 cm^{-1} is associated with the C-O-C stretching vibration of the polyester. The spectrum shows a significant decrease in the intensity of the peak at 1251 cm^{-1} in the HTP-2-1 resin, indicating that the amide bonds may have undergone cleavage. Given that the bond dissociation energies of the C-N and C-O bonds are 120–300 kJ/mol and 462 kJ/mol, respectively, it is likely that the C-N bond in the amide group undergoes cleavage first. This observation is consistent with the findings of Zhu et al., who attributed the lower protective performance of aliphatic polyurethanes to the cleavage of the C-N bond [22,23].

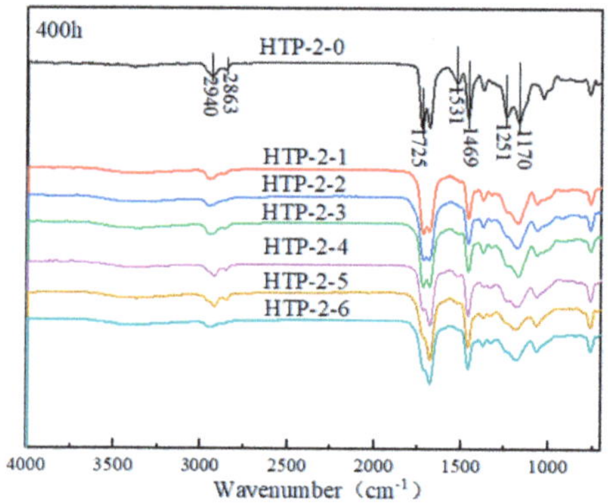

Figure 10. Infrared spectrum of the HTP-2-series resin after UV aging for 400 h.

4. Conclusions

Using two systems, neopentyl glycol-phthalic anhydride (HTP-1) and neopentyl glycol-hexahydrophthalic anhydride (HTP-2), the effects of various functionalized polyols on the properties of polyester-based polyurethane coatings were investigated. GPC results indicated that at low hydroxyl values, the molecular weight distribution was relatively narrow, while an increase in hydroxyl values resulted in a broader distribution. In the HTP-1 system, the introduction of dipentaerythritol led to a molecular weight distribution of as high as 12.66. DSC analysis revealed that dipentaerythritol in the HTP-1 system exhibited the highest glass transition temperature, reaching 40.19 °C.

Thermal analysis demonstrated that polyesters prepared with trimethylolpropane in the HTP-1 system exhibited inferior thermal stability, with the lowest mass retention of 86.49% at 300 °C. Coating performance tests showed that the coatings derived from the HTP-1 system exhibited excellent impact resistance, hardness, and adhesion, while flexibility decreased with increasing hydroxyl value. In contrast, coatings derived from the HTP-2 system showed poor adhesion at low hydroxyl values, improving from grade 1 to grade 0 as hydroxyl values increased, with a corresponding decrease in flexibility (from 1 mm to 2 mm) and lower hardness compared to the HTP-1 system.

Oil resistance tests revealed that coatings with low hydroxyl values were unable to resist oil penetration, with the HTP-2 system performing worse. However, as hydroxyl values increased, oil resistance significantly improved. Due to the steric hindrance of dipentaerythritol, which reduced crosslinking density, its oil resistance was comparatively poor. Regarding weather resistance, the HTP-1 series resins primarily exhibited degradation via the cleavage of $-CH_2$ groups, whereas the HTP-2-series resins predominantly exhibited C-N bond cleavage.

The research results demonstrate that the selection and proportion of polyfunctional polyols significantly influence the properties of polyester-based polyurethane coatings, including mechanical strength, flexibility, crosslinking density, and oil resistance. During polyester synthesis, it is essential to select appropriate polyols based on specific application requirements. For instance, to achieve high hardness in industrial protective coatings, such as those for machinery surfaces, ditrimethylolpropane (DTMP) can be employed as the primary polyol to enhance crosslinking density and coating durability. In the case of aerospace coatings, where superior oil resistance and excellent mechanical properties are critical, a combination of trimethylolpropane (TMP) and DTMP in high-hydroxyl-value systems can be utilized to achieve well-balanced performance optimization.

Author Contributions: Conceptualization, Z.Z.; Methodology, N.N.; Resources, Y.X.; Writing—original draft, Z.Z. All authors have read and agreed to the published version of the manuscript.

Funding: This research received no external funding.

Institutional Review Board Statement: Not applicable.

Informed Consent Statement: Not applicable.

Data Availability Statement: The original contributions presented in this study are included in the article. Further inquiries can be directed to the corresponding author.

Conflicts of Interest: The authors declare no conflicts of interest.

eferences

Akindoyo, J.O.; Beg, M.D.H.; Ghazali, S.; Islam, M.R.; Jeyaratnam, N.; Yuvaraj, A.R. Polyurethane types, synthesis and applications—A review. *RSC Adv.* **2016**, *6*, 114453–114482. [CrossRef]

Paraskar, P.M.; Prabhudesai, M.S.; Hatkar, V.M.; Kulkarni, R.D. Vegetable oil based polyurethane coatings—A sustainable approach: A review. *Prog. Org. Coat.* **2021**, *156*, 106267. [CrossRef]

Alves, T.R.; Pacheli Heitmann, A.; Ayres, E.; Coura, I.R.; de Souza, P.P.; de, O. Patricio, P.S. Waterborne polyurethane dispersions based on polypropylene glycol/polycarbonate diol: Evaluation of their use as wood protective coatings. *J. Appl. Polym. Sci.* **2023**, *140*, 1–11. [CrossRef]

Jian, H.; Wang, H.; Wen, M.; Shi, J.; Park, H.-J. High-performance flame-retardant, waterproof, and abrasion-resistant waterborne polyurethane-based transparent coatings for wood protection. *Ind. Crops Prod.* **2024**, *222*, 119804. [CrossRef]

Puyadena, M.; Etxeberria, I.; Ortega, G.; Sanz, O.; Agirre, A.; Mugica, A.; Matxinandiarena, E.; Barrio, A.; Gonzalez, A.; Irusta, L. Polyurethane/acrylic coatings for wood made hydrophobic by melamine addition. *Prog. Org. Coat.* **2024**, *186*, 108054. [CrossRef]

Cuevas, J.M.; Cobos, R.; German, L.; Sierra, B.; Laza, J.M.; Vilas-Vilela, J.L. Enhanced mar/scratch resistance in automotive clear coatings by modifying crosslinked polyurethane network with branched flexible oligomers. *Prog. Org. Coat.* **2022**, *163*, 106668. [CrossRef]

Lee, Y.-H.; Lee, S.-J.; Park, J.-W.; Kim, H.-J. Synthesis and properties of flexible polyester with urethane polyol for automotive pre-coated metals. *J. Adhes. Sci. Technol.* **2016**, *30*, 1537–1554. [CrossRef]

Oh, D.; Park, J.; Song, Y.K.; Noh, S.M.; Jung, H.W. Crosslinking characteristics of aziridine crosslinkers in polyurethane-based clearcoats for automotive applications. *J. Coat. Technol. Res.* **2024**, *21*, 1893–1906. [CrossRef]

Tan, H.; Wang, D.; Guo, Y. Investigation of graphene effect on the anti-corrosion behaviour of polyurethane coatings in sea-water. *Anti-Corros. Methods Mater.* **2019**, *66*, 853–860. [CrossRef]

Vinodhini, S.P.; Xavier, J.R. Effects of poly(o-phenylenediamine) functionalized SiC on the corrosion protection ability of neat polyurethane coating system in the marine environment. *Int. J. Polym. Anal. Charact.* **2023**, *28*, 113–138. [CrossRef]

Golling, F.E.; Pires, R.; Hecking, A.; Weikard, J.; Richter, F.; Danielmeier, K.; Dijkstra, D. Polyurethanes for coatings and adhesives—Chemistry and applications. *Polym. Int.* **2019**, *68*, 848–855. [CrossRef]

Liao, Y.-H.; Chen, Y.-C. Preparation and optimization of WPU dispersion from polyether/polyester polyols for film and coating applications. *J. Taiwan Inst. Chem. Eng.* **2023**, *145*, 104832. [CrossRef]

Jiang, W.; Jin, F.-L.; Park, S.-J. Synthesis of ditrimethylolpropane acrylate with low functionality for UV-curable coatings. *J. Ind. Eng. Chem.* **2012**, *18*, 1577–1581. [CrossRef]

14. Huang, S.; Xiao, J.; Zhu, Y.a.; Qu, J. Synthesis and properties of multiarm star hydroxyl-terminated polyesters for two-compor polyurethane coatings. *J. Coat. Technol. Res.* **2017**, *14*, 505–516. [CrossRef]
15. Haseebuddin, S.; Parmar, R.; Waghoo, G.; Ghosh, S.K. Study of hexafunctional polyol in high solids air-drying alkyd: Impro film performance. *Prog. Org. Coat.* **2009**, *64*, 446–453. [CrossRef]
16. Naik, R.B.; Ratna, D.; Singh, S.K. Synthesis and characterization of novel hyperbranched alkyd and isocyanate trimer based h solid polyurethane coatings. *Prog. Org. Coat.* **2014**, *77*, 369–379. [CrossRef]
17. Zhang, T.; Howell, B.A.; Smith, P.B. Thermal degradation of carboxy-terminal trimethylolpropane/adipic acid hyperbranc poly(ester)s. *J. Therm. Anal. Calorim.* **2015**, *122*, 1159–1166. [CrossRef]
18. Zhang, T.; Howell, B.A.; Smith, P.B. Thermal degradation of glycerol/adipic acid hyperbranched poly(ester)s containing ei hydroxyl or carboxyl end-groups. *J. Therm. Anal. Calorim.* **2015**, *122*, 1221–1229. [CrossRef]
19. Mayer, P.; Lubecki, M.; Stosiak, M.; Robakowska, M. Effects of surface preparation on the adhesion of UV-aged polyureth coatings. *Int. J. Adhes. Adhes.* **2022**, *117*, 103183. [CrossRef]
20. Yu, F.; Gao, J.; Liu, C.; Chen, Y.; Zhong, G.; Hodges, C.; Chen, M.; Zhang, H. Preparation and UV aging of nano-SiO_2/fluorina polyacrylate polyurethane hydrophobic composite coating. *Prog. Org. Coat.* **2020**, *141*, 105556. [CrossRef]
21. Rosu, D.; Rosu, L.; Cascaval, C.N. IR-change and yellowing of polyurethane as a result of UV irradiation. *Polym. Degrad. S* **2009**, *94*, 591–596. [CrossRef]
22. Mayer-Trzaskowska, P.; Robakowska, M.; Gierz, L.; Pach, J.; Mazur, E. Observation of the Effect of Aging on the Structu Changes of Polyurethane/Polyurea Coatings. *Polymers* **2024**, *16*, 23. [CrossRef] [PubMed]
23. Zhu, Y.; Xiong, J.; Tang, Y.; Zuo, Y. EIS study on failure process of two polyurethane composite coatings. *Prog. Org. Coat.* **2010**, 7–11. [CrossRef]

coatings

MDPI

rticle

Preparation of Antistatic Polyester Fiber via Layer-by-Layer Self-Assembly

ei Wang [1,2], Jialong Zhang [3], Yifan Liu [1,2], Mengyun Weng [1,2] and Yanchun Fu [1,2,*]

1 College of Furnishings and Industrial Design, Nanjing Forestry University, Nanjing 210037, China; wangwei1219@njfu.edu.cn (W.W.); 2241400155@njfu.edu.cn (Y.L.); wmy2180@njfu.edu.cn (M.W.)
2 Jiangsu Co-Innovation Center of Efficient Processing and Utilization of Forest Resources, Nanjing Forestry University, Nanjing 210037, China
3 College of Art and Design, Nanjing Forestry University, Nanjing 210037, China; 1274886861@njfu.edu.cn
* Correspondence: yanchunfu@njfu.edu.cn; Tel.: +86-25-8542-7793

Abstract: Polyester fibers tend to generate static electricity during the weaving and application processes, posing a threat to their production. Enhancing the water absorbency and electrical conductivity of polyester fibers themselves is an effective approach to improving their antistatic properties. In this study, multifunctional chitosan (CS), sodium phytate (SP), and Cu^{2+} were loaded on polyester fibers through layer-by-layer (LBL) self-assembly. The antistatic and water absorption capability of the modified polyester fibers was investigated by designing different process parameters combined with a surface resistance test and water contact angle tests. The antistatic property test results confirmed the positive effect of CS and Cu^{2+} on discharging electrostatic charge. Within a definite scope, with the increase in the number of assembly layers, assembly duration, and the concentration of the assembly substances, the wettability of the modified polyester fibers became more favorable and the antistatic effect became more remarkable.

Keywords: polyester fiber; antistatic property; wettability; layer-by-layer self-assembly

1. Introduction

In the present era of organic synthesis and the development of macromolecule polymers, an increasing number of highly practical polyester fibers with different specialties are being developed and put into production and use [1,2]. Polyester fibers are widely used in the textile industry for their superior mechanical properties, comfort, and breathing abilities, such as high tensile strength and modulus, moderate resilience, excellent thermal setting effects, good heat, light resistance, etc. [3,4]. Polyester fiber has become the fiber with the fastest development and the largest output among synthetic fibers due to the good controllability of the preparation process and excellent product performance [5,6]. However, the significant problem of static electricity generation has limited its production and application to a certain extent. Effective antistatic modification is urgently needed to expand the application range of polyester fiber [7–9]. Various polymeric modifiers are utilized for the permanent antistatic modification of synthetic fibers. Sano et al. designed copoly(amide-ethers)-containing ionic units as antistatic modifiers to form hygroscopic paths, which greatly reduce resistivity. Even after dyeing, the antistatic properties persisted. This is due to the fact that both hydrophilic and ionic groups had been effectively immobilized on the surface of polyester fibers [10,11].

Metal ion coordination modification is an effective method utilized to create functional fabrics [12–15]. This process forms a crosslinked structure via coordination between metal ions and the ligand functional groups present on the fabric's surface. In this way, the antistatic, antimicrobial, and other physicochemical properties of fabrics can be effectively improved for a wide range of demanding applications. Xu et al. employed phytic acid, seafoam, polyaspartic acid, and Fe^{3+} as raw materials and successfully prepared dense protective film coatings on the surface of cotton fabrics by using a layer-by-layer self-assembly method and spraying technique. This protective layer endowed cotton fabric with long-lasting flame-retardant properties [16]. Nabipour et al. deposited guanazole complexes containing silver and zinc ions on the surface of cotton fabrics by layer-by-layer self-assembly. The treated cotton fabrics showed excellent flame retardancy with a UL-94 V-0 rating in a vertical combustion test. At the same time, they also had excellent antimicrobial properties, and the silver guanazole-treated cotton fabrics also had good antifungal activity [17]. In recent years, there has been a growing interest in employing layer-by-layer (LBL) self-assembly technology for the antistatic modification of polyester fibers [18–20]. LBL self-assembly is a highly versatile and used technique for surface modification that involves the sequential adsorption of a polyanion and a polycation to form thin polymer films [21,22]. This technique offers a broad range of materials for constructing layer-by-layer assemblies, from basic polyelectrolytes and organic small molecules to advanced functional nanoparticles and even polyvalent metal ions [23–25]. While most research efforts have traditionally focused on synthetic polyelectrolytes, natural macromolecules such as chitosan, pectin, sodium alginate, and xanthan gum are also promising materials for constructing polyelectrolyte layers. These natural polyelectrolytes exhibit excellent water solubility and are more amenable to cross-linking with other substances, which enhances their adsorption properties and interface stability [26–28].

In this study, we developed a method for constructing an antistatic coating on the surface of polyester fibers via LBL self-assembled technology. The LBL self-assembly process involved the sequential deposition of chitosan, a cationic polyelectrolyte, sodium phytate, an anionic polyelectrolyte, and Cu^{2+} to form a composite coating on the polyester fiber surface. In search of a more feasible and efficient process for the preparation of antistatic polyester fibers, a series of multiple experiments were conducted with different numbers of assembled layers, assembly durations, and concentrations of assembled substances as variables.

2. Materials and Methods

2.1. Materials

Dust-free polyester fiber cloth specimen: 100% low-elastic polyester fiber (75D/36F, straight grain, weave, 110 g/m^2) was purchased from Bufan Industrial Co., Ltd. (Dongguan, China), with a smooth surface, firm sealing edge, and no obvious defects. Chitosan (degree of deacetylation $\geq$ 95%, viscosity 100–200 mPa·s), sodium phytate (99.0%), and copric chloride dihydrate ($CuCl_2 \cdot 2H_2O$, purity 99.9% on a metal basis) were supplied by Shanghai Machlin Biochemical Co., Ltd., Shanghai, China. Hydrochloric acid (36.5–38.0%) was purchased from Nanjing Chemical Reagent Co., Ltd., Nanjing, China. The distilled water utilized in the experiments was produced by a laboratory-grade ultrapure water system (model PLUS-E3-10TH, provided by Nanjing Yipu Yida Technology Development Co., Ltd., Nanjing, China).

2.2. Pre-Treatment of Polyester Fiber Substrate

The polyester fibers were made into samples with sizes of 100 mm × 100 mm × 10 mm, 40 mm × 40 mm × 1 mm, and 15 mm × 15 mm × 1 mm, respectively, which were used for antistatic and surface wettability experiments and characterization. The substrates of the polyester fibers were cleaned twice with deionized water until they were free of dust impurities. Then, they were placed in an oven at 60 °C for drying for 0.5 h, removed, and stored in a desiccator for standby.

2.3. Preparation of the Self-Assembly Solution

The positively charged CS solution was prepared by dissolving CS powder in distilled water. Subsequently, dilute hydrochloric acid was added dropwise into to CS solution and stirred at a constant rate until the pH of the solution was 3. The anionic SP solution was prepared by dissolving SP powder in distilled water, stirring at a constant speed, and gradually adding dilute hydrochloric acid until the pH of the solution was 3. For the Cu^{2+} solution, $CuCl_2{\cdot}2H_2O$ powder was dissolved in distilled water while being continuously stirred.

2.4. Formation of Antistatic Coatings on Polyester Fibers via Layer-by-Layer Self-Assembly

In this experiment, the LBL self-assembly technique was employed, with the mutual adsorption of CS, SP, and Cu^{2+} resulting in the formation of antistatic coatings on polyester fibers. First, the polyester fiber specimens were impregnated in the CS solution and immersed for a period of time at room temperature and atmospheric pressure. As the surface of the polyester fibers had a negative charge after the surface activation treatment in the early stage, the positive and negative charges [29] were attracted to each other, and the surface of the polyester fibers was successfully loaded with CS polycation layers. After the impregnation, the specimens were removed, cleaned with distilled water, and dried in a drying oven at 50 °C for 0.5 h. At this time, the polyester fibers possessed a positive surface charge. Then, they were impregnated in the SP solution and immersed for a period of time. After the impregnation, the specimen was removed, cleaned with distilled water, and dried in a drying oven at 50°C for 0.5 h. At this time, the surface of the polyester fibers had a negative charge. Subsequently, the specimens continued to be impregnated in the solution of $CuCl_2$ for some time. Then, the samples were rinsed with distilled water and dried in the oven at 50 °C for another 30 min. One layer of SP and one layer of Cu^{2+} were considered to comprise a single deposition cycle. After that, the SP and the Cu^{2+} layers were loaded several times using cyclic loading to ensure the performance effect. The preparation process is schematically shown in Figure 1.

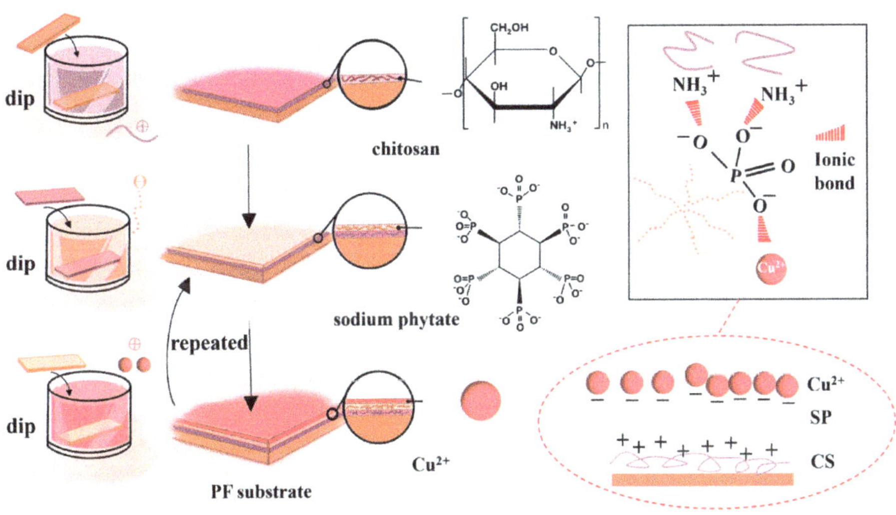

Figure 1. Scheme of fabricating CS/SP/Cu^{2+} coatings on polyester fiber surface by LBL self-assembly.

The concentrations of polyelectrolyte and Cu^{2+} solutions, the soaking time, and the number of layers assembled were taken as variables in the growth control experiments. The original polyester fiber was taken as the control group. Multiple experiments were set up to explore the optimal process conditions or process range for the surface performance improvement in polyester fiber. The concentration of all the assembly substances was set as 1, 5, and 10 g/L. The soaking time was set as 30, 60, and 90 min. Moreover, the number of SP/Cu^{2+} loading layers were set as 1, 3, and 6. When exploring the effect of a particular variable, all other conditions were kept identical.

2.5. Characterization and Measurements

The surface morphology and elemental composition of the polyester fiber coatings were characterized using scanning electron microscopy (SEM, Quanta 400 FEG, FEI Inc., Eindhoven, The Netherlands) equipped with an energy-dispersive X-ray spectroscopy (EDX) system. Each sample was coated with a 5–10 nm Au layer prior to SEM imaging. A Fourier transform infrared spectroscopy (FT-IR) analysis of the samples was performed using an ALPHA II infrared spectrometer (ALPHA II, Bruker, Fällanden, Switzerland) in the scanning region of 4000~600 cm^{-1}.

2.6. Surface Resistance Test

The surface resistance (R) of the samples (size: $100 \times 100 \times 10$ mm^3) was measured at a temperature of 22 °C and 51% humidity using a handheld surface resistivity meter (VC385, Victorlong Instruments, Shenzhen, China). The average value was calculated after every three point measurements. The antistatic effect of the samples was evaluated and analyzed according to the standard AATCC 76-2018 [30] and GB/T 12703.4-2010 [31].

2.7. Water Absorption Performance Test

The water contact angle was measured using a contact angle system (CA, CAST2.0 CA, Solon Information Technology Co., Ltd., Shanghai, China) at room temperature. Distilled water droplets of 5 μL each were employed for the contact angle analysis. To evaluate the change in the contact angle over time, the test solution was continuously dispensed onto the polyester fiber for one minute. Three distinct locations on the sample were measured, and an average was determined. Additionally, the contact angle of the unmodified fiber was measured to serve as a comparative reference.

3. Results and Discussion

3.1. Surface Morphology and Structure Analysis

The surface morphology of the polyester fibers before and after modification was investigated by SEM, as shown in Figure 2a–e. As observed, the surface of the untreated polyester fibers was smooth but with some granular impurities. After a 90 min deposition of CS (1 g/L) and SP (1 g/L), the fibers displayed a more rugged appearance, indicating the formation of a coating wrapping around the surface of the polyester fibers (Figure 2b,c). When a layer of $CuCl_2$ (1 g/L) was deposited on the polyester fiber surface for 90 min, merely a small quantity of particle-like impurities were present (Figure 2d). As it was further deposited up to three layers, the number of particles seemed to pile up and agglomerate. However, the distribution remained uneven (Figure 2e). Following the assembly of six layers, a thick and compact coating appeared on the entire surface of the polyester fiber (Figure 2f,g). As shown in Figure 2h, there are N, P, and Cu elements from CS, SP, and $CuCl_2$ on the coated polyester fiber. The results demonstrate that the CS/SP/Cu^{2+} coating was successfully formed.

The FT-IR spectra of the above specimen are also shown in Figure 3. The peaks present across the wavelength of 1720 cm^{-1} to 720 cm^{-1} were related to the cyclic group in the polyester fiber structure. The characteristic peaks of the polyester fiber became weaker after the application of the CS/SP/Cu^{2+} coating, indicating that the polyester fiber surface was covered by the self-assembled layer. The diminished peaks in the infrared spectrum provide less information because the typical absorption peaks of the polyester fiber were very strong in this wavenumber range. Additionally, the absorption peaks in the range of 1600–1700 cm^{-1}, corresponding to -OH groups, showed a shift, which may be due to the formation of hydrogen bonds [32,33].

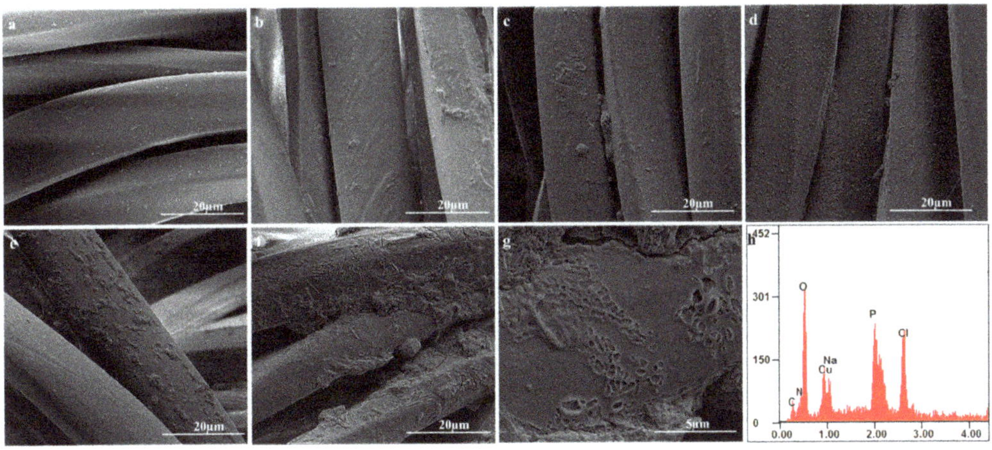

Figure 2. SEM images of (**a**) original polyester fiber, (**b**) CS-coated polyester fiber, (**c**) CS/SP-coated polyester fiber, (**d**) 1 layer of CS/SP/Cu^{2+}-coated polyester fiber, (**e**) 3 layers of CS/SP/Cu^{2+}-coated polyester fiber, 6 layers of CS/SP/Cu^{2+}-coated polyester fiber at (**f**) 5000× and (**g**) 20,000× magnification, and (**h**) EDS analysis of CS/SP/Cu^{2+}-coated polyester fiber.

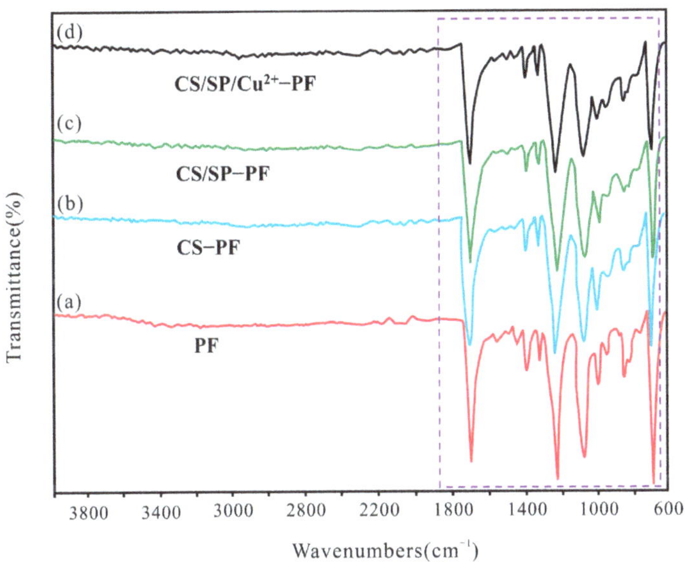

Figure 3. FT-IR spectroscopy of (**a**) original, (**b**) CS-coated, (**c**) CS/SP-coated, and (**d**) CS/SP/Cu^{2+}-coated polyester fiber samples.

3.2. Analysis of Surface Resistance

The data concerning the effects of varying numbers of assembly layers, assembly durations, and concentrations of the assembly substances influencing the surface resistance of the polyester fibers are summarized in Table 1. The surface resistance of the original polyester fiber exceeded the measurement range of the resistance tester, surpassing 10^{12} Ω, and thus classifying it as an insulating material. The polyester fibers with a single layer of CS loading exhibited a reduction in surface resistance, which further decreased as the CS concentration increased, ultimately reaching as low as 10^{10} Ω. This indicated that the loading of CS improved the antistatic property of the polyester fiber. A higher CS concentration means more positively charged ions, creating more conductive channels on the fiber surface that facilitate charge migration and dissipation [34]. However, the addition of SP seemed to have no significant impact on the antistatic property of the polyester fiber. The incorporation of conductive Cu^{2+} led to a marked decrease in surface resistance, with the reduction continuing as the concentration increased. Under the identical concentration (10 g/L) and number of assembly layers (six layers), the surface resistance of the polyester fiber treated for a 30 min immersion was 10^8 Ω, which met the basic antistatic requirement for polyester fibers. Extending the immersion time to 90 min further reduced the surface resistance to as low as 10^6 Ω. It can be observed that a longer immersion duration of the polyelectrolyte and Cu^{2+} on the polyester fiber leads to a greater reduction in surface resistance and a more pronounced antistatic effect. A similar tendency was also observed with variations in the number of loading layers. An increase in the number of loading layers corresponded to a decrease in surface resistance and an improvement in its antistatic property. The antistatic durability of the modified polyester fiber was further evaluated by washing experiments. According to GB/T 12490-2014 [35], the washing solution was added to a water bath containing stainless steel beads. A sample (10 g/L, 90 min soaking, and six layers) was taken and stirred in the water bath at 40 °C for 30 min. This process was recorded as one washing time. The test sample still maintained an antistatic surface resistivity of about 10^8 Ω after being washed 10 times.

Table 1. Variation in surface resistance of polyester fibers with different numbers of assembly layers, assembly durations, and concentrations of assembly substances.

LBL Self-Assembly Structure	Concentration (g/L)			Impregnation Time(min)	Layers	Surface Resistance Value (Ω)
	CS	SP	$CuCl_2$			
original	/	/	/	/	/	outrange
CS	1	/	/	90	1	10^{12}
CS	5	/	/	90	1	10^{11}
CS	10	/	/	90	1	10^{10}
CS/SP	1	1	/	90	6	10^{11}
CS/SP	5	5	/	90	6	10^{10}
CS/SP	10	10	/	90	6	10^{10}
CS/SP/Cu^{2+}	1	1	1	90	6	10^9
CS/SP/Cu^{2+}	5	5	5	90	6	10^8
CS/SP/Cu^{2+}	10	10	10	90	6	10^6
CS/SP/Cu^{2+}	10	10	10	30	6	10^8
CS/SP/Cu^{2+}	10	10	10	60	6	10^7
CS/SP/Cu^{2+}	10	10	10	90	1	10^8
CS/SP/Cu^{2+}	10	10	10	90	3	10^7
CS/SP/Cu^{2+} after washing 10 times	10	10	10	90	6	10^8

3.3. *Analysis of Surface Wettability*

The water absorption data of the modified polyester fibers are revealed through the water contact angle test (Figure 4). An analysis of the experimental data (Figure 4a) shows that the water contact angle of the original polyester fiber was 136°, with no obvious change observed within 60 s. This lack of change is attributed to the synthetic nature of polyester fibers, which feature a dense molecular structure that inhibits water molecule infiltration and absorption. In contrast, the initial water contact angle of the sample loaded with one layer of CS was 133°, but decreased to 52° after 60 s, indicating a notable improvement in its water absorption property. This enhancement is due to the presence of a large number of hydrophilic groups, such as hydroxyl and amino groups, in the CS structure [36]. After loading SP, there was no significant change in the water absorption properties. This result aligns with the surface resistance test findings, indicating that SP did not improve the antistatic effect and mainly served as a bridge to connect CS and Cu^{2+}. However, the introduction of Cu^{2+} led to a dramatic reduction in the water contact angle from 130° to 0° within 20 s, demonstrating a remarkable improvement in water absorption compared to the previous samples. The CS/SP/Cu^{2+} coating altered the surface morphology and microstructure of the polyester fiber, facilitating water molecule penetration and absorption [34]. Water absorption performance under different assembly conditions was analyzed, revealing a trend similar to that observed in the surface resistance test. Specifically, a greater number of loading layers within a certain range improved the water absorption effect (Figure 4b). In this experiment, the most effective water absorption was achieved when both the nanoparticle and polyelectrolyte concentrations were 10 g/L (Figure 4c). At this concentration, water was absorbed almost instantly upon contact. Additionally, the water absorption of polyester fibers increased with prolonged treatment time (Figure 4d). It is evident that all polyester fibers treated with CS/SP/Cu^{2+} exhibited water droplet absorption within 5 s. This finding indicates that the treatment method significantly enhances the water absorption capacity of the polyester fibers, which, in turn, facilitates charge conduction and dissipation, thereby improving the antistatic performance.

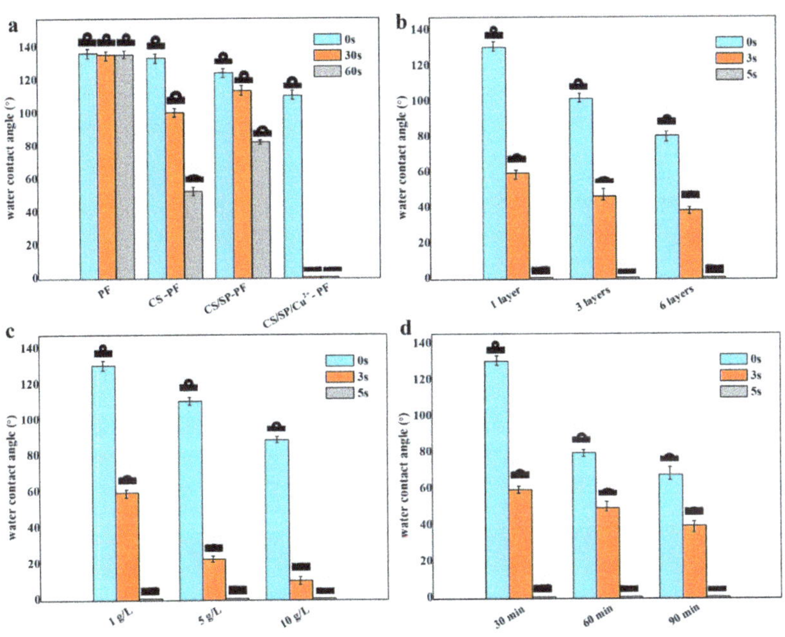

Figure 4. The water contact angle data of (**a**) different LBL self-assembly structures, (**b**) the numbers of assembly layers, (**c**) the concentrations, and (**d**) the assembly durations of the assembly substances affecting the wettability of the polyester fibers.

4. Conclusions

This study successfully achieved efficient antistatic CS/SP/Cu^{2+} coatings on polyester fiber surfaces via the LBL self-assembly method. Under various self-assembly conditions, the surface resistance values of the modified polyester fiber samples ranged from 10^{12} to 10^6 Ω, meeting the requirements of antistatic materials and demonstrating a good antistatic effect. Generally, both Cu^{2+} and CS were effective in enhancing the antistatic property. As the number of assembly layers, assembly durations, and the concentration of assembly substances increased, the antistatic performance improved correspondingly. Based on a comprehensive analysis of the test results, the optimal conditions for achieving a superior antistatic property in this experiment are an assembly time of 90 min, an assembly cycle of six layers, and a concentration of 10 g/L. Further washing experiments showed that the modified samples still maintained an antistatic surface resistivity of about 10^8 Ω after washing 10 times and good antistatic durability.

Author Contributions: Conceptualization, W.W. and J.Z.; methodology, W.W., M.W. and Y.F.; software, Y.L.; validation, W.W., J.Z. and Y.L.; formal analysis, W.W., J.Z. and M.W.; investigation, Y.L.; resources, W.W.; data curation, W.W.; writing—original draft preparation, Y.F., Y.L. and J.Z.; writing—review and editing, W.W. and Y.F.; visualization, J.Z.; supervision, W.W.; project administration, W.W. and Y.F. All authors have read and agreed to the published version of the manuscript.

Funding: This research was funded by the Natural Science Foundation of Jiangsu Province (grant number BK20220426) and Natural Science Research of Jiangsu Higher Education Institutions of China (grant number 22KJB220004).

Institutional Review Board Statement: Not applicable.

Informed Consent Statement: Not applicable.

Data Availability Statement: The original contributions presented in the study are included in the article, further inquiries can be directed to the corresponding author.

Acknowledgments: The authors thank the Institute of Chemical Industry of Forest Products for providing invaluable technical assistance in the chemistry laboratory.

Conflicts of Interest: The authors declare no conflicts of interest.

References

1. Li, P.; Wang, Q.Z.; Wang, B.; Liu, Y.; Xu, Y.; Liu, Y.; Zhu, P. Blending alginate fibers with polyester fibers for flame-retardant filli materials: Thermal decomposition behaviors and fire performance. *Polym. Degrad. Stab.* **2021**, *193*, 109470. [CrossRef]
2. Fan, Y.; Shen, J.; Xu, H. Synthesis and dilute aqueous solution properties of cationic antistatic surfactant functionalized w hydroxyl and ether groups. *Tenside Surfactants Deterg.* **2023**, *60*, 64–73. [CrossRef]
3. Lv, J.; Zhou, Q.; Zhi, T.; Gao, D.; Wang, C. Environmentally friendly surface modification of polyethylene terephthalate (PI fabric by low-temperature oxygen plasma and carboxymethyl chitosan. *J. Clean. Prod.* **2016**, *118*, 187–196. [CrossRef]
4. Khoddami, A.; Avinc, O.; Ghahremanzadeh, F. Improvement in poly (lactic acid) fabric performance via hydrophilic coati *Prog. Org. Coat.* **2011**, *72*, 299–304. [CrossRef]
5. Xu, Y.; Wu, N.; Wei, Q.; Chu, L. Surface Functionalisation of Polyester Nonwoven Fabrics by Sputter Coating of Titanium Dioxi *Polym. Polym. Compos.* **2009**, *17*, 347–351. [CrossRef]
6. Li, S.; Huang, S.; Xu, F.; Xiao, H.; Zhang, F.; Zhang, G. Preparing polyester/carbon multifunctional fabrics by phosphoric a carbonization. *Cellulose* **2019**, *26*, 8907–8917. [CrossRef]
7. Kan, C.W. Evaluating antistatic performance of plasma-treated polyester. *Fibers Polym.* **2007**, *8*, 629–634. [CrossRef]
8. An, W.; Ma, J.; Xu, Q.; Fan, Q. Flame retardant, antistatic cotton fabrics crafted by layer-by-layer assembly. *Cellulose* **2020**, 8457–8469. [CrossRef]
9. Su, M.; Chen, X.; Zhang, L.; Min, J. Synthesis of Active Graphene with Para-Ester on Cotton Fabrics for Antistatic Properti *Nanomaterials* **2020**, *10*, 1147. [CrossRef]
10. Sano, Y.; Lee, C.; Kimura, Y.; Saegusa, T. A novel spin-finishing method for antistatic modification of polyester fiber. Surfa reaction onto amine-containing copoly(amide-ethers) blended with polyester fiber. *Die Angew. Makromol. Chem. Appl. Macron Chem. Phys.* **1997**, *246*, 109–123. [CrossRef]
11. Sano, Y.; Konda, M.; Lee, C.; Kimura, Y.; Saegusa, T. A facile antistatic modification of polyester fiber based on ion-exchan reaction of sulfonate-modified polyester and various cationic surfactants. *Die Angew. Makromol. Chem. Appl. Macromol. Che Phys.* **1997**, *251*, 181–191. [CrossRef]
12. Lindgren, J.; Persson, P.; Öhman, L.O. Interactions of calcium (II), copper (II) and aluminium (III) ions with two chemical modified wood fibres. *Nord. Pulp Pap. Res. J.* **2001**, *16*, 225–234. [CrossRef]
13. Xing, X.; Yang, H.; Tao, M.; Zhang, W. An overwhelmingly selective colorimetric sensor for Ag^+ using a simple modifi polyacrylonitrile fiber. *J. Hazard. Mater.* **2015**, *297*, 207–216. [CrossRef]
14. Gritsch, L.; Lovell, C.; Goldmann, W.H.; Boccaccini, A.R. Fabrication and characterization of copper (II)-chitosan complexes antibiotic-free antibacterial biomaterial. *Carbohydr. Polym.* **2018**, *179*, 370–378. [CrossRef] [PubMed]
15. Zhao, S.; Wang, F.; Zhou, R.; Liu, P.; Xiong, Q.; Zhang, W.; Ye, X.; Gao, H. Fabrication of recyclable Fe^{3+} chelated aminat polypropylene fiber for efficient clean-up of phosphate wastewater. *Front. Chem. Sci. Eng.* **2023**, *17*, 840–852. [CrossRef]
16. Xu, T.; Qian, D.; Hu, Y.; Zhu, Y.; Zhong, Y.; Zhang, L.; Xu, H.; Mao, Z. Assembled hybrid films based on sepiolite, phytic aci polyaspartic acid and Fe^{3+} for flame-retardant cotton fabric. *J. Polym. Eng.* **2022**, *42*, 744–754. [CrossRef]

27. Nabipour, H.; Wang, X.; Rahman, M.Z.; Song, L.; Hu, Y. An environmentally friendly approach to fabricating flame retardant, antibacterial and antifungal cotton fabrics via self-assembly of guanazole-metal complex. *J. Clean. Prod.* **2020**, *273*, 122832. [CrossRef]
28. Aslam, S.; Hussain, T.; Ashraf, M.; Tabassum, M.; Rehman, A.; Iqbal, K.; Javid, A. Multifunctional Finishing of Cotton Fabric. *Autex Res. J.* **2019**, *19*, 191–200. [CrossRef]
29. Pan, Y.; Liang, Q.; Song, L.; Zhao, H. Fabrication of layer-by-layer self-assembled coating modified cotton fabric with flame retardancy and hydrophobicity based on sepiolite. *Polym.-Plast. Technol. Mater.* **2021**, *60*, 1368–1376. [CrossRef]
30. Chen, X.; Ding, F.; Zhang, S.; Liu, Y.; Hou, X.; Ren, X. Flame-retardant, antibacterial and hydrophobic multifunctional coatings on cotton fabrics via layer-by-layer self-assembly. *Cellulose* **2023**, *30*, 6679–6694. [CrossRef]
31. Wang, X.; Romero, M.Q.; Zhang, X.Q.; Wang, R.; Wang, D.Y. Intumescent multilayer hybrid coating for flame retardant cotton fabrics based on layer-by-layer assembly and sol–gel process. *RSC Adv.* **2015**, *5*, 10647–10655. [CrossRef]
32. Iler, R.K. Multilayers of colloidal particles. *J. Colloid Interface Sci.* **1966**, *21*, 569–594. [CrossRef]
33. Decher, G.; Hong, J.D. Buildup of Ultrathin Multilayer Films by a Self-Assembly Process. I. Consecutive Adsorption of Anionic and Cationic Bipolar Amphiphiles on Charged Surfaces. *Makromol. Chem. Macromol. Symp.* **1991**, *46*, 321–327. [CrossRef]
34. Jeonghwan, K.; Arcadio, S.; Jungwoo, C.; Dowoo, N.; Arman, B.; Bruggen, B.V. Embedding TiO_2 nanoparticles versus surface coating by layer-by-layer deposition on nanoporous polymeric films. *Microporous Mesoporous Mater.* **2013**, *173*, 121–128. [CrossRef]
35. Wang, J.; Wang, F.; Xu, Z.; Wang, Y.; Dong, S. Surface plasmon resonance and electrochemistry characterization of layer-by-layer self-assembled DNA and Zr^{4+} thin films, and their interaction with cytochrome C. *Talanta* **2007**, *74*, 104–109. [CrossRef] [PubMed]
36. Iamphaojeen, Y.; Siriphannon, P. Immobilization of zinc oxide nanoparticles on cotton fabrics using poly 4-styrenesulfonic acid polyelectrolyte. *Int. J. Mater. Res.* **2012**, *103*, 643–647. [CrossRef]
37. Yang, J.M.; Tsai, R.Z.; Hsu, C.C. Protein adsorption on polyanion/polycation layer-by-layer assembled polyelectrolyte films. *Colloids Surf. B Biointerfaces* **2016**, *142*, 98. [CrossRef] [PubMed]
38. Janesch, J.; Czabany, I.; Hansmann, C.; Mautne, A.; Rosenau, T.; Altmutter, W.G. Transparent layer-by-layer coatings based on biopolymers and CeO_2 to protect wood from UV light. *Prog. Org. Coat.* **2020**, *138*, 105409. [CrossRef]
39. Teli, M.; Prasad, N.; Vyas, U. Electrokinetic properties of modified polyester fibers. *J. Appl. Polym. Sci.* **1993**, *50*, 449–457. [CrossRef]
40. *AATCC 76-2018(E2019)*; Test Method for Electrical Surface Resistivity of Fabrics. American Association of Textile Chemists and Colorists: Research Triangle Park, NC, USA, 2019.
41. *GB/T 12703.4-2010*; Textile-Evaluation for Electrostatic Properties—Part 4: Resistivity. Standards Press of China: Beijing, China, 2010.
42. Xie, M.; Zhang, F.; Peng, H.; Zhang, Y.; Li, Y.; Xu, Y.; Xie, J. Layer-by-layer modification of magnetic graphene oxide by chitosan and sodium alginate with enhanced dispersibility for targeted drug delivery and photothermal therapy. *Colloids Surf. B Biointerfaces* **2019**, *176*, 462–470. [CrossRef]
43. Zhao, Y.; Hu, J.; Hu, X.; Zhu, F.; Su, J.; Han, J. A novel strategy for fabrication of antistatic and antibacterial fabric via layer-by-layer self-assembly. *Surf. Coat. Technol.* **2023**, *453*, 129143. [CrossRef]
44. Eom, S. Using Chitosan as an Antistatic Finish for Polyester Fabric. *AATCC Rev.* **2001**, *1*, 57.
45. *GB/T 12490-2014*; Textiles-Tests for Colour Fastness-Colour Fastness to Domestic and Commercial Laundering. Standards Press of China: Beijing, China, 2014.
46. Abdel-Halima, E.; Abdel-Mohdya, F.; Al-Deyabb, S.; El-Newehy, E. Chitosan and monochlorotriazinyl–cyclodextrin finishes improve antistatic properties of cotton/polyester blend and polyester fabrics. *Carbohydr. Polym.* **2010**, *82*, 202–208. [CrossRef]

coatings

MDPI

Article

Synthesis and Performance Evaluation of Bio-Sourced PO3G Ester Plasticizer in L-Polylactic Acid Thin Films

Hongzhao Tian, Yiyi Zhu, Xinyue Kong, Jiao Long, Wan Zhao, Dengbang Jiang * and Mingwei Yuan *

Green Preparation Technology of Biobased Materials National & Local Joint Engineering Research Center, Yunnan Minzu University, Kunming 650500, China; 18225395951@163.com (H.T.); 18214491715@163.com (J.L.); ymuzhaowan@163.com (W.Z.)

* Correspondence: 041814@ymu.edu.cn (D.J.); ymuyuan2024@163.com (M.Y.)

Abstract: This study aims to synthesize and evaluate the properties of bio-sourced poly(1,3-propanediol) laurate (PO3G-LA) as a plasticizer in the fabrication of poly(1,3-propanediol) laurate/L-polylactic acid (PO3G-LA/PLLA) thin films. Utilizing an esterification reaction between poly(1,3-propanediol) (PO3G) and lauric acid (LA), PO3G-LA is synthesized and incorporated into PLLA films via solution casting. Results demonstrate that PO3G-LA exhibits excellent compatibility with PLLA, markedly enhancing the toughness and slightly improving the thermal stability of the resulting films. Specifically, the addition of 20% PO3G-LA increases the elongation at the break of the films to 190%, indicating enhanced flexibility. Accelerated degradation experiments conducted at 60 °C revealed that the presence of PO3G-LA under neutral conditions had minimal impact on the degradation rate of the film samples. Conversely, variations in the PO3G-LA content of the films when exposed to acidic and alkaline conditions were found to influence their degradation rates.

Keywords: bio-sourced plasticizer; poly(1,3-propanediol) laurate; L-polylactic acid (PLLA); thin film materials

1. Introduction

Polylactic acid (PLA) stands out among bio-based polyesters for its excellent thermal processability, optical properties, high modulus, environmental sustainability, and biocompatibility, making it a promising alternative to traditional plastics [1,2]. PLA finds extensive applications in the medical field, serving as an implant material, drug delivery system, and tissue engineering scaffold [3]. Its utility extends to commodity packaging, where its biodegradability and minimal environmental footprint offer a compelling substitute for traditional plastics [4]. Moreover, PLA's potential in textiles is under exploration, driven by its biodegradability and capacity to mitigate the environmental footprint of textile manufacturing [5]. However, the inherent brittleness and low elongation at the break of PLA materials limit their broader applications [6,7].

Common modification strategies to mitigate these limitations include chemical copolymerization and physical blending [8–10]. Physical blending modification is presently the prevailing approach in industrial production [11]. An established technique involves incorporating plasticizers, which can diminish intermolecular forces among PLA molecules, thereby enhancing the material's flexibility and processability [12,13]. Additives such as citrate esters [14], Ethyl triacetate (TAC) [15], and poly(1,2-propanediol) (PPG) have demonstrated efficacy in bolstering PLA's ductility and flexibility [16]. Buong et al. [17] found that blending a small amount of polyethylene glycol (PEG)-200 with PLLA yielded no significant toughening effect, but the toughness improved noticeably when the PEG-200 content exceeded 10 wt% [18]. Chu et al. [19] explored the toughening impact of varying N-ethyl-p-toluenesulfonamide (N-PTSA) concentrations on PLA. Their findings revealed enhanced motility and flexibility of PLA molecular chains upon N-PTSA addition, alongside improved interfacial compatibility and reduced melt processing viscosity, thus leading to cost savings. Lin et al. [20] combined PCL, TBC, and sodium dodecylbenzenesulphonate (SDBS)-modified multi-walled carbon nanotubes (s-MWCNTs) with PLA, synergistically enhancing PLA toughness through PCL flexibility and TBC's plasticizer effect, while s-MWCNTs acted as reinforcement. Hao et al. [21] utilized poly(diethylene glycol adipate) (PDEGA) as a plasticizer, finding that low molecular weight PDEGA (L-PDEGA) improved PLA's elongation at break, while high molecular weight PDEGA (H-PDEGA) enhanced its impact strength. These studies underscore the efficacy of physical blending and specific additive incorporation as strategies to enhance the mechanical properties of PLA, thereby broadening its application potential.

Poly(1,3-propanediol) (PO3G), a novel fully bio-derived polyether developed by DuPont, exhibits superior compliance, mechanical properties, and thermal stability attributed to its elongated molecular chain. PO3G is synthesized through the condensation of 1,3-propanediol. 1,3-Propanediol can be produced via a bio-based method using biomass such as sugarcane or corn as feedstock through microbial fermentation [22–24]. Therefore, PO3G is a potential high-quality green plasticizer derived entirely from renewable sources [25,26]. PO3G has similar structural units to PEG and PPG, suggesting its potential as a plasticizer for polylactic acid (PLLA). Nonetheless, the terminal hydroxyl group in PO3G was anticipated to expedite the degradation of PLLA. In response, this study pursued the esterification of PO3G with lauric acid (LA), leading to the synthesis of a new PLLA plasticizer, poly(1,3-propanediol) laurate (PO3G-LA). PO3G-LA utilizes the renewable and biodegradable nature of its principal raw material, PO3G, and aligns with PLLA in environmental sustainability, presenting a green and natural copolymer that minimizes environmental impacts typically associated with plasticizers. The introduction of PO3G-LA into PLLA films significantly enhances their elongation at break, thermal stability, and crystallinity, positioning PO3G-LA as an innovative green plasticizer and a viable alternative to conventional plasticizers.

2. Experimental

2.1. *Experimental Reagents*

Materials used in this study include L-polylactic acid (PLLA, 4032D, Nature Works, Minnetonka, MN, USA), poly(1,3-propanediol) (PO3G, Mn = 2300, DuPont, Wilmington, NC, USA), lauric acid (LA, Chengdu Kelong Chemical Reagent Factory, Chengdu, China), 1,4-dioxane (Adamas Reagent Co., Shanghai, China), molybdenum trioxide (MoO3, Adamas Reagent Company, Shanghai, China), and dichloromethane (DCM, Damao Chemical Reagent Factory, Tianjin, China).

2.2. *Experimental Apparatus*

The equipment utilized encompasses a collector-type constant temperature heating magnetic stirrer (DF-101S, Gongyi Yuhua Instrument Co., Ltd., Gongyi, China); an electric blast drying oven (GZX-9240MBE, Bo Xun Industrial Co., Ltd., medical equipment factory, Shanghai, China); an electronic balance (FA2004, Koshihira Scientific Instrument Co., Ltd., Shanghai, China); a 1HNMR (400 MHz Bruker Avance II, Bruker, Billerica, MA, USA); a Fourier Transform Infrared Spectrum Analyzer (IS10, Nicolet Instrument Co., Ltd., Madison, WI, USA); an SEM (NOVA NANOSEM 450, FEI, Hillsboro, OR, USA); a Microcomputer Controlled Universal Tensile Tester (CMT4104, Chuangcheng Zhijia Science and Technology Co., Ltd., Beijing, China); a Thermogravimetric Analyzer (STA449F3, NETZSCH, Selb, Germany); and a Differential Scanning Calorimeter (2414Polyma, NETZSCH, Bavaria, Germany).

2.3. *Preparation of PO3G-LA*

In a 250 mL three-necked flask, 143.75 g of PO3G and 32 g of LA, along with 3.125 g of molybdenum trioxide, were combined. The flask was initially purged with dry nitrogen at room temperature for 15 min, then heated to 160 °C with uniform stirring under a nitrogen stream for a 6 h reaction period. Upon cooling to room temperature, an appropriate amount of dichloromethane was added to dissolve the reaction mixture. The catalyst molybdenum trioxide and any unreacted lauric acid were removed through filtration, performed five times to ensure thorough separation. Subsequently, excess water was eliminated via distillation under reduced pressure, yielding 152.85 g of a light yellow viscous liquid, identified as PO3G-LA. The synthesis route of PO3G-LA is illustrated in Figure 1.

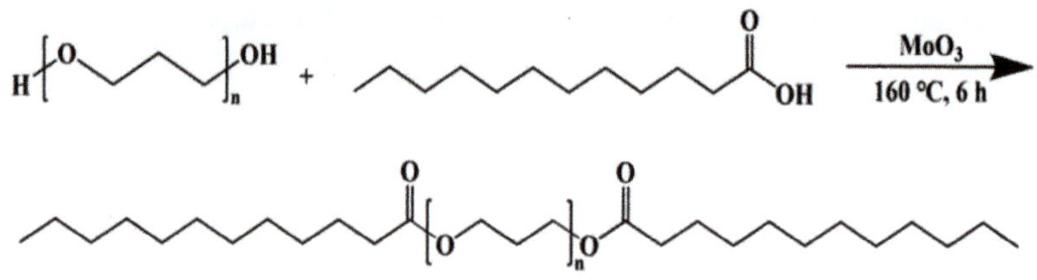

Figure 1. Synthesis route of PO3G-LA.

2.4. PO3G-LA/PLLA Composite Film Blending Experiment with Different Ratios

For film preparation, a casting solution with a 20% mass fraction was created using dioxane as the solvent and PO3G-LA and PLLA as solutes. The mass fractions of PO3G-LA to PLLA varied at 0%, 5%, 10%, 15%, 20%, and 25%, resulting in composite films designated as PLLA, 5PO3G-LA/PLLA, 10PO3G-LA/PLLA, 15PO3G-LA/PLLA, 20PO3G-LA/PLLA, and 25PO3G-LA/PLLA. The composite material was dissolved in a three-necked flask to form a homogeneous, colorless, viscous liquid. This solution was then transferred to a conical flask.

A clean glass plate was prepared, and a measured amount of the solution was evenly spread across the glass plate's surface using a 500 μm film scraper to form a thin film. The glass plate was left to air-dry in a fume hood for 6 h, followed by drying in a 60 °C oven for an additional 6 h. After oven drying, the film-coated glass plate was left in the oven at 60 °C for 24 more hours to ensure complete drying. The film was then removed, sealed, and stored for further analysis. The schematic diagram of PO3G-LA/PLLA film preparation is depicted in Figure 2.

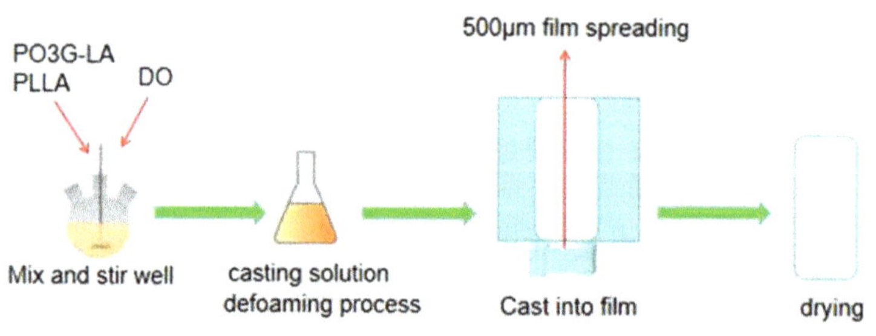

Figure 2. Schematic diagram of PO3G-LA/PLLA film preparation.

2.5. Performance Testing and Structural Characterization

Nuclear magnetic resonance hydrogen spectroscopy test. Approximately 5 mg of PO3G, LA, and PO3G-LA were weighed using an analytical balance, dissolved in about 0.05 g of $CDCL_3$, and then transferred to an NMR tube for analysis.

Fourier transform infrared spectroscopy (FTIR) Characterization. One mg each of PO3G, LA, PO3G-LA, as well as samples of PLLA and PO3G-LA/PLLA films with varying PO3G-LA contents (5%, 10%, 15%, 20%, 25%), were dissolved in a suitable amount of dichloromethane. These solutions were then applied to KBr plates—prepared by repeated milling and pressing of KBr powders—dried of solvent and scanned in the wavenumber range of 4000–400 cm^{-1} using FTIR.

Thermogravimetric analysis (TGA). TGA was employed to determine the thermal decomposition temperatures of the PO3G-LA/PLLA film samples with different PO3G-LA ratios. The experiments were conducted under a nitrogen gas flow, heating at a rate of 10 °C/min within the temperature range of 25–600 °C.

Differential scanning calorimetry (DSC) characterization. DSC analyzed the thermal behavior of the PO3G-LA/PLLA films across a temperature range of −40 °C to 190 °C, with a heating rate of 10 °C/min. The crystallinity of the films with various ratios was calculated using the following equation:

$$X_c = \frac{\Delta H_m}{\Delta H_0 \times \omega_{PLLA}} \times 100\%,$$

where ΔH_m is the enthalpy of melting of the film sample, ΔH_0 is the enthalpy of crystallization for fully crystallized PLLA under ideal conditions (93.6 J/g), and ω is the mass fraction of PLLA in the film sample.

Mechanical performance test. The tensile strength and elongation at the break of the film samples were measured using an electronic universal testing machine. Specimens were prepared as rectangles measuring 50 × 20 mm and tested at a tensile rate of 10 mm/min, with the experiment conducted five times for each ratio of film samples to obtain average values.

Surface morphology characterization. The surface morphology of the films was examined under a scanning electron microscope (SEM). Five dry specimens were fixed on a sample stage using conductive adhesive and sputter-coated with gold for 30 s to enhance conductivity. Images of the surface structures were captured at consistent magnification levels using SEM.

Biodegradability test. Twenty-one square samples, each measuring 20 × 20 mm^2, were prepared from PO3G-LA/PLLA films with varying PO3G-LA contents and subsequently dried under vacuum at 65 °C for 20 h to achieve a constant weight, which was recorded as m_0. These samples were then placed into twenty-one sample bottles, each receiving 20 mL of a pre-prepared buffer solution at pH levels of 3.0, 7.0, and 11.0. The bottles were securely sealed and incubated in an oven set at 60 °C to undergo degradation experiments, with the buffer solutions being refreshed every 48 h over a 21-day degradation cycle.

3. Results and Analysis

3.1. Nuclear Magnetic Resonance Map Hydrogen Characterization of PO3G-LA

The 1H-NMR spectrum of LA reveals characteristic chemical shifts (Figure 3): 2.3 ppm for -$\underline{CH_2}$-(C=O)OH, 1.2 ppm and 1.32 ppm for -CH_2-$\underline{CH_2}$-CH_2-, and 1.6 ppm and 0.75 ppm for $\underline{CH_3}$-$\underline{CH_2}$-. In the PO3G spectrum, the peaks at 3.55 ppm, 3.4 ppm, and 1.8 ppm correspond to -$\underline{CH_2}$-OH, -CH_2-$\underline{CH_2}$-O, and -CH_2-$\underline{CH_2}$-CH_2-, respectively. The integral areas of these peaks at 3.55 ppm, 3.4 ppm, and 1.8 ppm were 1, 19.67, and 39.11, respectively, allowing for the calculation of PO3G's average degree of polymerization to be about 20, with a molecular weight (M_n) of 2300. The PO3G-LA profile exhibited a shift in the chemical shifts of 3.55 ppm and 3.6 ppm to a higher chemical shift at 4.1 ppm, with the 4.1 ppm peak corresponding to $\underline{CH_2}$-O(C=O)-. This shift indicates the completion of the esterification reaction between the hydroxyl group on PO3G and lauric acid, signifying the formation of PO3G-LA.

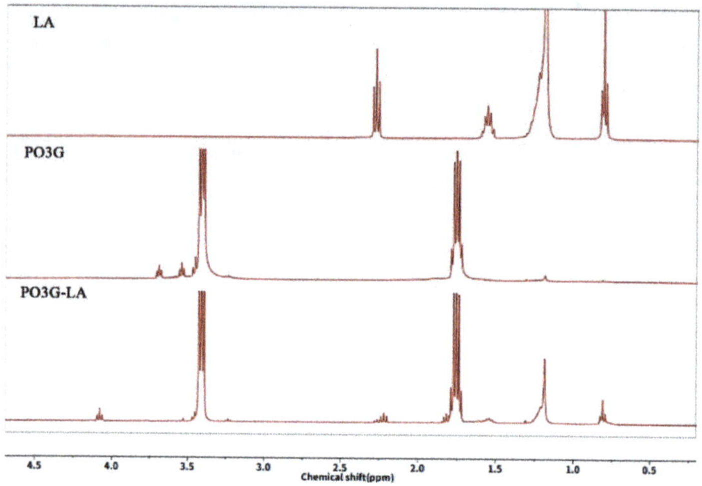

Figure 3. 1H-NMR characterization of LA, PO3G, and PO3G-LAA.

3.2. Fourier Infrared Spectroscopic Characterization of PO3G-LA, PO3G-LA/PLLA Films

Figure 4 illustrates the FTIR analysis, where the peak at 2924 cm^{-1} in the LA spectrum is identified as the stretching vibration of C-H, and the peak at 1687 cm^{-1} corresponds to the C=O stretching vibration. In the PO3G spectrum, the 1121 cm^{-1} peak is attributed to the C-O stretching vibration. Notably, in the PO3G-LA spectra, the C=O stretching vibration shifts to 1732 cm^{-1}, altering the carboxylate absorption peak seen at 1687 cm^{-1} in LA's spectrum, confirming the occurrence of an esterification reaction.

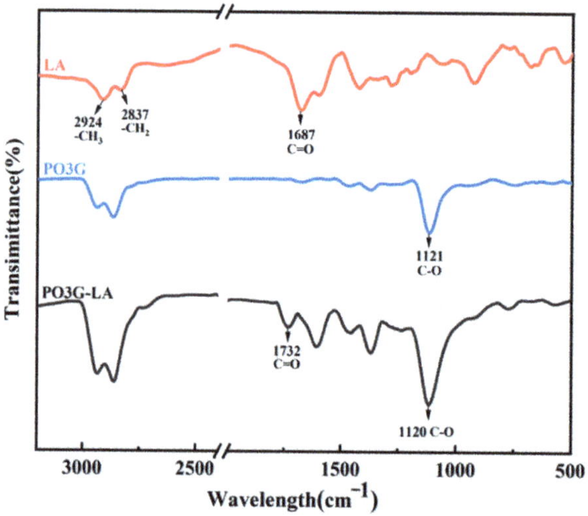

Figure 4. FTIR spectra of PO3G-LA, PO3G, and LA.

Figure 5 displays the FTIR spectra of PO3G-LA/PLLA composite films and those of pure PLLA films. The spectra of the composite films closely mirror those of PLLA without the emergence of new characteristic peaks or the disappearance of existing ones, aside from variations in peak intensity. The peak at 1758 cm^{-1} aligns with the C-O stretching vibration in both PLLA and PO3G-LA, while the peaks at 1367 cm^{-1} and 1457 cm^{-1} are proximate to the C-H stretching vibrations in -CH3 and -CH2-. Additionally, the peak at

1090 cm^{-1} represents the C-O stretching vibration. These spectral characteristics indicate the successful synthesis of PO3G-LA and its incorporation into the PO3G-LA/PLLA films.

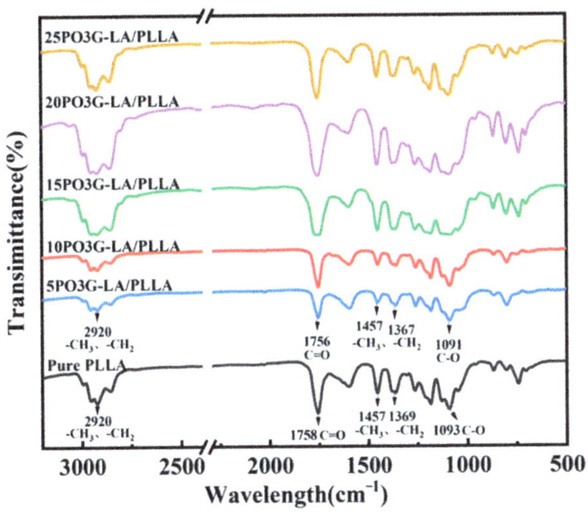

Figure 5. FTIR spectra of PO3G-LA/PLLA films with different PO3G-LA contents.

3.3. *Analysis of Thermal Stability and Crystallization Properties of PO3G-LA/PLLA Films*

Figure 6 presents the thermogravimetric (TGA) and derivative thermogravimetry (DTG) analyses for PO3G-LA/PLLA composite films with varying concentrations of PO3G-LA. Incorporating PO3G-LA into PLLA films not only slightly elevates the temperature required to reach a 50% weight loss, as seen in the TGA curves, but also diminishes the overall thermal decomposition. This enhancement in thermal stability is further evidenced by the DTG curves, which demonstrate a consistent decline in the peak thermal loss rate with an increase in PO3G-LA concentration. Notably, the inclusion of 5% PO3G-LA shows an increase in the thermal loss rate for the composite films compared to pure PLLA. However, when the PO3G-LA concentration is raised to 10% or beyond, a notable decrease in the thermal loss rate is observed, indicating an improvement in thermal stability.

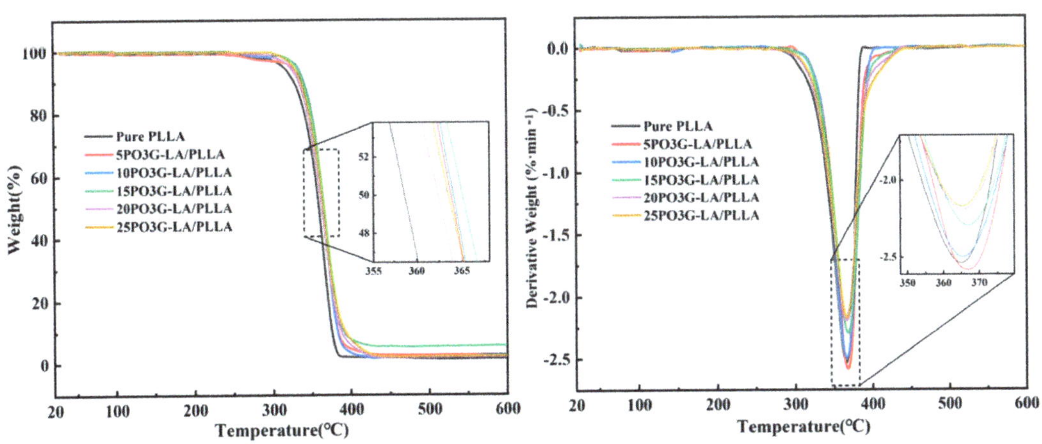

Figure 6. Thermogravimetric (TGA) (**left**) and derivative thermogravimetry (DTG) (**right**) curves of PO3G-LA/PLLA composite films with different PO3G-LA contents.

This phenomenon can be attributed to the plasticizing effect of PO3G-LA, which increases the mobility of the PLLA chains, thereby facilitating a more orderly thermal decomposition process. The presence of PO3G-LA likely aids in dispersing thermal energy more evenly throughout the polymer matrix, which in turn elevates the degradation temperature. This mechanism is supported by literature indicating that plasticizers can improve the thermal stability of biopolymers by enhancing the mobility of polymer chains [27].

Furthermore, studies have shown that the incorporation of bio-based plasticizers into biopolymers not only enhances their flexibility but also can slightly increase their thermal stability due to improved energy dissipation and thermal resistance [28]. These references corroborate our findings that the addition of PO3G-LA, particularly at concentrations of 10% or higher, contributes to the enhanced thermal stability of PLLA films, making it a promising plasticizer for improving the thermal properties of bio-based polymers.

Figure 7 showcases the Differential Scanning Calorimetry (DSC) curves for PO3G-LA/PLLA composite films across a range of PO3G-LA concentrations, with corresponding quantitative data detailed in Table 1. The analysis of Table 1 elucidates that the incorporation of PO3G-LA not only lowers the glass transition temperature (Tg) of the PLLA films but also influences the crystallinity of the composites in a noteworthy manner. Specifically, a nuanced variation in Tg is observed, which initially decreases with the addition of PO3G-LA, reaching a nadir at 50.2 °C with a 20% PO3G-LA content, indicative of the highest level of compatibility between PO3G-LA and PLLA at this or lower concentrations. An upward shift in Tg with a 25% PO3G-LA inclusion signals a slight reduction in compatibility. Moreover, the crystallinity of the films undergoes a significant alteration upon PO3G-LA addition. The pure PLLA film exhibits a crystallinity of 35.76%, whereas the incorporation of 5%, 10%, 15%, 20%, and 25% PO3G-LA enhances the crystallinity to 39.64%, 37.95%, 43.19%, 45.33%, and 40.62%, respectively. This increment in crystallinity with PO3G-LA addition underscores its role as an efficacious nucleating agent, promoting the orderly arrangement of PLLA chains and thus enhancing the crystallinity of the composite films.

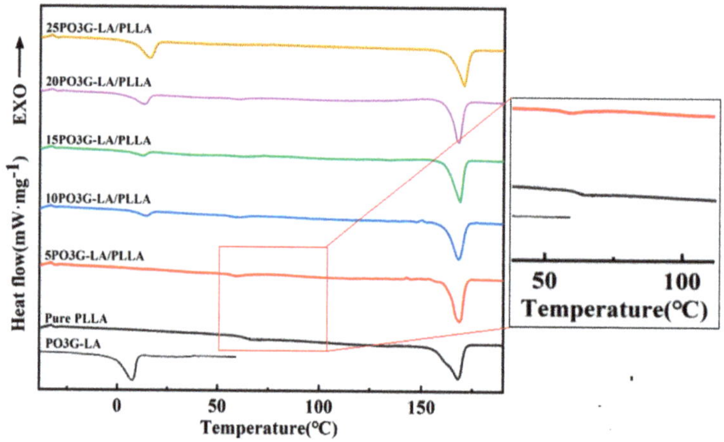

Figure 7. DSC curves of PO3G-LA/PLLA films with different PO3G-LA contents.

Table 1. DSC data of PO3G-LA/PLLA films with varying PO3G-LA contents.

Sample	*Tg* (°C)	*Tm* (°C)	Δ*Hm* (J/g)	*Xc* (%)
Pure PLLA	60.4	166.8	33.48	35.76
5PO3G–LA/PLLA	54.9	168.3	35.25	39.64
10PO3G–LA/PLLA	55.8	168.0	31.97	37.95
15PO3G–LA/PLLA	52.4	167.2	34.36	43.19
20PO3G–LA/PLLA	50.2	167.0	33.95	45.33
25PO3G–LA/PLLA	56.1	167.9	28.52	40.62

It is important to note the absence of a cold crystallization peak in the DSC curves, which is typically observed in polymers with significant amorphous regions undergoing heating. This absence can be attributed to the films' high degree of crystallinity achieved during the solution casting process, possibly nearing the material's crystalline saturation. This high initial crystallinity limits the availability of amorphous regions that could recrystallize upon heating, hence the non-observation of cold crystallization peaks in our DSC analysis. This phenomenon, coupled with the observed increases in crystallinity upon PO3G-LA addition, suggests an enhanced ordering and densification of the polymer matrix, contributing to the thermal and structural properties of the PLLA composites.

3.4. Analysis of Mechanical Properties of PO3G-LA/PLLA Films

Analysis from Table 2 reveals that the pure PLLA film exhibits an elongation at a break of only 5.71%. However, the introduction of PO3G-LA into PLLA films significantly enhances their ductility, with elongation at break values reaching 77%, 91%, 139%, and 190% for PO3G-LA additions of 5%, 10%, 15%, 20%, and 25%, respectively. This improvement demonstrates a progressive increase in the elongation at break for PO3G-LA contents ranging from 0% to 20%. A subsequent addition of PO3G-LA to 25% results in a reduction of elongation at break to 99%. Figure 8 illustrates that with increasing PO3G-LA content, the elongation at break initially increases and then decreases while the tensile strength of the films gradually declines. This observed pattern suggests that excessive PO3G-LA concentrations may lead to its aggregation, negatively impacting the material's toughening effect on PLLA films. Notably, the tensile strength of the composite films decreases marginally with PO3G-LA additions between 5% and 25%.

Table 2. Elongation at break and tensile strength of PO3G-LA/PLLA composite films with different PO3G-LA contents.

Sample	Elongation at Break/%	Tensile Strength/MPa
Pure PLLA	5.71 ± 1.93	42.46 ± 1.32
5PO3G–LA/PLLA	77.33 ± 6.27	30.24 ± 0.49
10PO3G–LA/PLLA	91.54 ± 2.40	29.18 ± 0.43
15PO3G–LA/PLLA	139.38 ± 5.00	27.21 ± 0.77
20PO3G–LA/PLLA	190.88 ± 4.84	26.46 ± 1.69
25PO3G–LA/PLLA	99.95 ± 6.62	24.33 ± 1.18

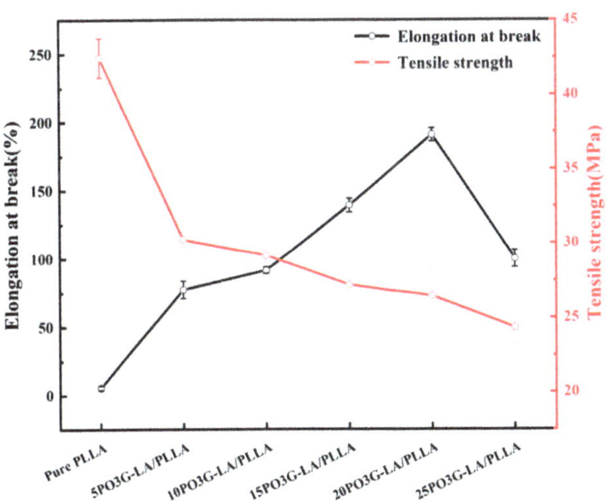

Figure 8. Elongation at break and tensile strength of PO3G-LA/PLLA composite films with different PO3G-LA contents.

The molecular structure of PO3G-LA, characterized by ether bonds along the main chain, facilitates more flexible rotation around these bonds, thereby enhancing chain flexibility. This property suggests that PO3G-LA holds promise as an effective plasticizer for PLLA, capable of improving its mechanical properties by enhancing ductility without significantly compromising tensile strength.

3.5. PO3G-LA/PLLA Thin Films Indicating Morphology Analysis

Figure 9 illustrates the surface morphologies of PO3G-LA/PLLA composite films with varying PO3G-LA contents. The films, prepared in different ratios, were laid on a paper sheet marked with "FILM". As observed in Figure 8, the macroscopic morphology of the composite films exhibited minimal change across different PO3G-LA concentrations, maintaining high transparency.

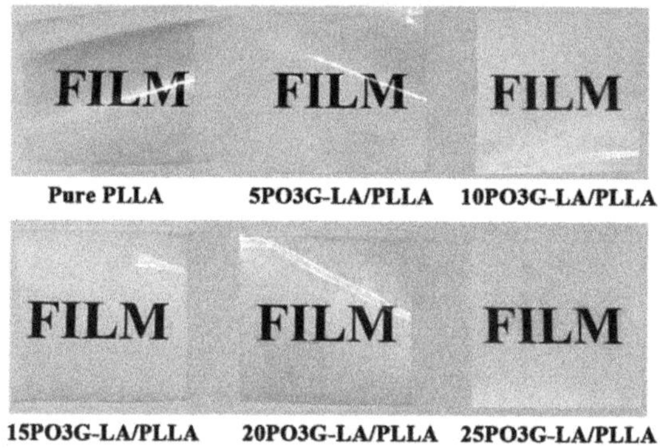

Figure 9. Surface morphology of PO3G-LA/PLLA composite films with different PO3G-LA contents.

3.6. Microscopic Morphology Analysis of PO3G-LA/PLLA Films

Figure 10 provides a detailed view of the microscopic morphologies of PO3G-LA/PLLA composite films with varying PO3G-LA concentrations. The surfaces of these films are observed to be uniformly smooth and flat across all samples. However, an increase in PO3G-LA content leads to a rougher surface texture and the appearance of bulges without any evident phase separation or surface material precipitation. This observation indicates excellent compatibility between PO3G-LA and PLLA, suggesting that PO3G-LA integrates well into the PLLA matrix without causing adverse effects on the film's uniformity.

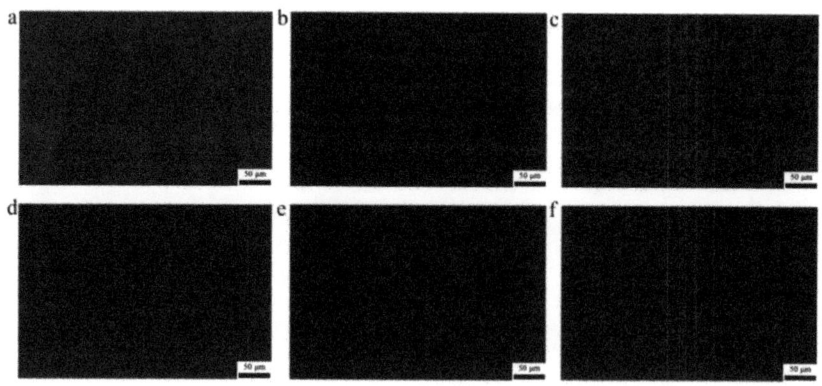

Figure 10. Micro-morphology of PO3G-LA/PLLA composite films with different PO3G-LA contents: (**a**) Pure PLLA (**b**) 5% PO3G-LA/PLLA (**c**) 10% PO3G-LA/PLLA (**d**) 15% PO3G-LA/PLLA (**e**) 20% PO3G-LA/PLLA (**f**) 25% PO3G-LA/PLLA.

Figure 11 shows the tensile fracture morphologies of PO3G-LA/PLLA composite films, contrasting the effects of different PO3G-LA concentrations on fracture behavior. Pure PLLA films exhibit relatively flat fracture surfaces, indicative of brittle failure. The introduction of PO3G-LA transforms these surfaces, making them rougher and adorned with burrs resulting from the material's stretching. This morphological change is emblematic of toughness fractures, underscoring that PO3G-LA incorporation significantly enhances the toughness of PLLA films.

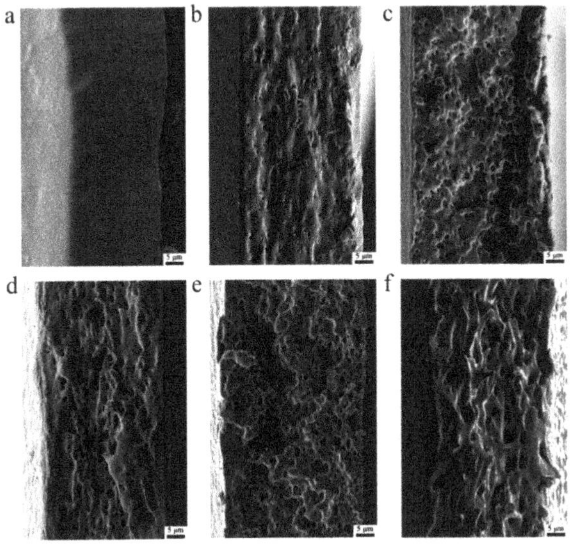

Figure 11. Tensile cross-sectional morphology of PO3G-LA/PLLA films with different PO3G-LA contents: (**a**) Pure PLLA (**b**) 5% PO3G-LA/PLLA (**c**) 10% PO3G-LA/PLLA (**d**) 15% PO3G-LA/PLLA (**e**) 20% PO3G-LA/PLLA (**f**) 25% PO3G-LA/PLLA.

Furthermore, numerous holes were noted along the tensile direction in the film sections, primarily attributable to the dispersed phase. This dispersed phase causes the PO3G-LA to deform along the direction of tension, transforming from spherical to elliptical shapes. The tips of the ellipses act as energy-absorbing sites, which during detachment from the substrate, absorb significant energy, leading to the formation of holes in the direction of tension.

3.7. Biodegradability Testing of PO3G-LA/PLLA Films

The degradation rate of polyester film materials can be significantly accelerated under acidic or alkaline conditions compared to the conventional degradation process in soil or compost, as evidenced in previous studies [29–31]. These environments have been shown to exert a profound influence on the degradation of biodegradable polymeric materials. Moreover, conducting degradation experiments at temperatures near the glass transition temperature (Tg) of PLLA enhances the mobility of macromolecular chains, thus expediting the degradation process and increasing experimental efficiency [32–34]. This mechanism primarily involves ester bond hydrolysis, resulting in the gradual breakdown of both PLLA and PO3G-LA/PLLA films, as observed in various research investigations [29,35,36].

Figure 12 illustrates that under neutral conditions (pH = 7), the addition of PO3G-LA had minimal impact on the film's degradation rate. However, under acidic conditions, the degradation rates of PO3G-LA/PLLA composite films with 5% to 25% PO3G-LA content were all lower compared to pure PLLA films. Among these, the 5PO3G-LA/PLLA films showed higher degradation rates than those containing 10PO3G-LA/PLLA, 15PO3G-

LA/PLLA, 20PO3G-LA/PLLA, and 25PO3G-LA/PLLA, films, and the degradation rates of the latter three ratios of composite films showed no significant differences. This trend could be attributed to the fact that PO3G-LA addition enhances the crystallinity of the PLLA films. As the content of PO3G-LA increases, the crystallinity of the PLLA films also increases, which, in turn, slows down the degradation rate.

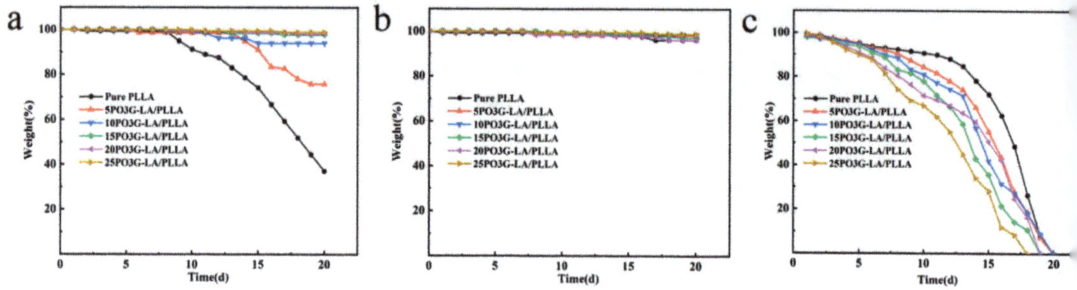

Figure 12. pH degradation curves of PO3G-LA/PLLA composite films with different PO3G-LA contents: (**a**) pH = 3, (**b**) pH = 7, (**c**) pH = 11.

Figure 12 also demonstrates that the films degrade more rapidly under alkaline conditions (pH 11) compared to neutral and acidic environments. Initially, the degradation rate of the composite film is primarily influenced by the PO3G-LA content, as PO3G-LA hydrolyzes more swiftly in alkaline conditions than in acidic ones, as indicated by references [37,38]. An increase in PO3G-LA content enhances the penetration of the solution into the film's interior, resulting in an accelerated degradation rate. Subsequent degradation rates are affected by multiple factors. The 25PO3G-LA/PLLA composite films were completely degraded by day 18, whereas the 20PO3G-LA/PLLA and 15PO3G-LA/PLLA films reached full degradation by day 19. The 10PO3G-LA/PLLA, 5PO3G-LA/PLLA, and pure PLLA films were all completely degraded by day 20. This variation in degradation times may be associated with the infiltration of small-molecule plasticizers and alterations in the crystallinity of the PLLA films.

4. Conclusions

In our research, we developed an eco-friendly, non-toxic ester plasticizer, PO3G-LA, from PO3G and LA, significantly enhancing PLA toughness. The toughness and thermal stability of PLLA films improved notably with the addition of PO3G-LA, particularly at higher concentrations. This improvement highlights PO3G-LA's potential as a sustainable plasticizer in toughening PLLA. Future research should explore its broader applications in sustainable materials development and investigate its behavior under varied environmental conditions to fully harness its ecological benefits.

Author Contributions: Methodology, H.T., Y.Z. and M.Y.; Validation, H.T., Y.Z. and X.K.; Formal analysis, Y.Z. and W.Z.; Data curation, H.T., X.K., J.L. and W.Z.; Writing—original draft, H.T.; Visualization, J.L.; Supervision, D.J.; Project administration, D.J.; Funding acquisition, M.Y. All authors have read and agreed to the published version of the manuscript.

Funding: This research was funded by the National Natural Science Foundation of China (52163013).

Institutional Review Board Statement: Not applicable.

Informed Consent Statement: Not applicable.

Data Availability Statement: Data are contained within the article.

Conflicts of Interest: The author declares no conflicts of interest.

eferences

Jem, K.J.; Tan, B. The development and challenges of poly (lactic acid) and poly (glycolic acid). *Adv. Ind. Eng. Polym. Res.* **2020**, *3*, 60–70. [CrossRef]

Pang, X.; Zhuang, X.; Tang, Z.; Chen, X. Polylactic acid (PLA): Research, development and industrialization. *Biotechnol. J.* **2010**, *5*, 1125–1136. [CrossRef]

Tyler, B.; Gullotti, D.; Mangraviti, A.; Utsuki, T.; Brem, H. Polylactic acid (PLA) controlled delivery carriers for biomedical applications. *Adv. Drug Deliv. Rev.* **2016**, *107*, 163–175. [CrossRef]

Mohamad, N.; Mazlan, M.M.; Tawakkal, I.S.M.A.; Talib, R.A.; Kian, L.K.; Fouad, H.; Jawaid, M. Development of active agents filled polylactic acid films for food packaging application. *Int. J. Biol. Macromol.* **2020**, *163*, 1451–1457. [CrossRef]

Guruprasad, R.; Prasad, G.K.; Prabu, G.; Raj, S.; Patil, P.G. Low-stress mechanical properties and fabric hand of cotton and pol-ylactic acid fibre blended knitted fabrics. *Indian J. Fibre Text. Res.* **2018**, *43*, 381–384.

Muller, J.; González-Martínez, C.; Chiralt, A. Combination of poly(lactic) acid and starch for biodegradable food packaging. *Materials* **2017**, *10*, 952. [CrossRef]

Malek, N.S.; Faizuwan, M.; Khusaimi, Z.; Bonnia, N.N.; Rusop, M.; Asli, N.A. Preparation and Characterization of Biodegradable Polylactic Acid (PLA) Film for Food Packaging Application: A Review. *J. Phys. Conf. Ser.* **2021**, *1892*, 012037. [CrossRef]

Auras, R.; Harte, B.; Selke, S. An Overview of Polylactides as PackagingMaterials. *Macromol. Biosci.* **2004**, *4*, 835–864. [CrossRef]

Yang, Y.; Zhang, L.; Xiong, Z.; Tang, Z.; Zhang, R.; Zhu, J. Research progress in the heat resistance, toughening and filling modification of PLA. *Sci. China Chem.* **2016**, *59*, 1355–1368. [CrossRef]

0. Ding, Y.; Lu, B.; Wang, P.; Wang, G.; Ji, J. PLA-PBAT-PLA tri-block copolymers: Effective compatibilizers for promotion of the mechanical and rheological properties of PLA/PBAT blends. *Polym. Degrad. Stab.* **2018**, *147*, 41–48. [CrossRef]

1. Akindoyo, J.O.; Beg, M.D.; Ghazali, S.; Heim, H.P.; Feldmann, M. Impact modified PLA-hydroxyapatite composites–Thermo-mechanical properties. *Compos. Part A Appl. Sci. Manuf.* **2018**, *107*, 326–333. [CrossRef]

2. Puthumana, M.; Santhana Gopala Krishnan, P.; Nayak, S.K. Chemical modifications of PLA through copolymerization. *Int. J. Polym. Anal. Charact.* **2020**, *25*, 634–648. [CrossRef]

3. Labrecque, L.V.; Kumar, R.A.; Gross, R.A.; McCarthy, S.P.; Davé, V. Citrate esters as plasticizers for poly(lactic acid). *J. Appl. Polym. Sci.* **1997**, *66*, 1507–1513. [CrossRef]

4. Mazitova, A.K.; Aminova, G.K.; Vikhareva, I.N. Designing of green plasticizers and assessment of the effectiveness of their use. *Polymers* **2021**, *13*, 1761. [CrossRef]

5. Nagarajan, V.; Mohanty, A.K.; Misra, M. Perspective on Polylactic Acid (PLA) based Sustainable Materials for Durable Ap-plications: Focus on Toughness and Heat Resistance. *ACS Sustain. Chem. Eng.* **2016**, *4*, 2899–2916. [CrossRef]

6. Krishnan, S.; Pandey, P.; Mohanty, S.; Nayak, S.K. Toughening of Polylactic Acid: An Overview of Research Progress. *Polym. Technol. Eng.* **2016**, *55*, 1623–1652. [CrossRef]

7. Xie, D.; Zhao, Y.; Li, Y.; LaChance, A.M.; Lai, J.; Sun, L.; Chen, J. Rheological, Thermal, and Degradation Properties of PLA/PPG Blends. *Materials* **2019**, *12*, 3519. [CrossRef]

8. Chieng, B.W.; Ibrahim, N.A.; Yunus, W.M.Z.W.; Hussein, M.Z. Plasticized poly (lactic acid) with low molecular weight poly (ethylene glycol): Mechanical, thermal, and morphology properties. *J. Appl. Polym. Sci.* **2013**, *130*, 4576–4580. [CrossRef]

9. Chu, W.; Yan, S.; Cai, Z.; Yang, S.; Hu, X.; Yin, X. Poly(lactic acid) toughen by aromatic sulfonamide toward balanced rheology processing and mechanical properties. *Mater. Res. Express* **2022**, *9*, 065307. [CrossRef]

20. Lin, G.; Wang, J.; Lv, N.; Wang, H.; Wang, H.; Yu, B.; Liu, S.; Liang, D.; Wang, Q.; Yu, K. Simultaneously Strengthening and Toughening Poly (lactic acid) by Co-Additions of Poly (ε-caprolactone), Tributyl Citrate Plasticizer and Functionalized Multiwall Carbon Nanotube. *Sci. Adv. Mater.* **2020**, *12*, 950–957. [CrossRef]

21. Jia, Z.; Tan, J.; Han, C.; Yang, Y.; Dong, L. Poly(ethylene glycol-*co*-propylene glycol) as a macromolecular plasticizing agent for polylactide: Thermomechanical properties and aging. *J. Appl. Polym. Sci.* **2009**, *114*, 1105–1117. [CrossRef]

22. Vo, A.D.; Cui, W.J.; McAuley, K.B. An Improved PO3G Model–Accounting for Cyclic Oligomers. *Macromol. Theory Simul.* **2020**, *29*, 2000023. [CrossRef]

23. Zhang, C.; Luan, H.; Wang, G. A novel thermosensitive triblock copolymer from 100% renewably sourced poly(trimethylene ether) glycol. *J. Appl. Polym. Sci.* **2018**, *135*, 46112. [CrossRef]

24. Yang, J.-X.; Qian, H.-J.; Gong, Z.; Lu, Z.-Y.; Cui, S.-X. Stretching Elasticity and Flexibility of Single Polyformaldehyde Chain. *Chin. J. Polym. Sci.* **2022**, *40*, 333–337. [CrossRef]

25. Maiza, M.; Benaniba, M.T.; Massardier-Nageotte, V. Plasticizing effects of citrate esters on properties of poly(lactic acid). *J. Polym. Eng.* **2016**, *36*, 371–380. [CrossRef]

26. Butt, J.; Oxford, P.; Sadeghi-Esfahlani, S.; Ghorabian, M.; Shirvani, H. Hybrid Manufacturing and Mechanical Characterization of Cu/PLA Composites. *Arab. J. Sci. Eng.* **2020**, *45*, 9339–9356. [CrossRef]

27. Okpuwhara, R.O.; Oboirien, B.O.; Sadiku, E.R. The lanolin-based oil plasticized polylactide: Thermal and chemical character-istics. *Polym. Eng. Sci.* **2022**, *62*, 1571. [CrossRef]

28. Rozaki, N.Z.; Gan, S.-N.; Ang, D.T.-C. Environmentally Friendly Oil-Modified Polyesters as Polymeric Plasticizers for Poly(vinyl chloride). *J. Polym. Environ.* **2017**, *25*, 286–295. [CrossRef]

29. Elsawy, M.A.; Kim, K.-H.; Park, J.-W.; Deep, A. Hydrolytic degradation of polylactic acid (PLA) and its composites. *Renew. Sustain. Energy Rev.* **2017**, *79*, 1346–1352. [CrossRef]

30. Ebadi-Dehaghani, H.; Barikani, M.; Borhani, S.; Bolvardi, B.; Khonakdar, H.A.; Jafari, S.H.; Aarabi, A. Biodegradation ar hy-drolysis studies on polypropylene/polylactide/organo-clay nanocomposites. *Polym. Bull.* **2016**, *73*, 3287–3304. [CrossRef]
31. Kumar, S.; Singh, S.; Senapati, S.; Singh, A.P.; Ray, B.; Maiti, P. Controlled drug release through regulated biodegradation poly(lactic acid) using inorganic salts. *Int. J. Biol. Macromol.* **2017**, *104*, 487–497. [CrossRef]
32. Hottle, T.A.; Agüero, M.L.; Bilec, M.M.; Landis, A.E. Alkaline amendment for the enhancement of compost degradation f polylactic acid biopolymer products. *Compos. Sci. Util.* **2016**, *24*, 159–173. [CrossRef]
33. Gómez, E.F.; Michel, F.C. Biodegradability of conventional and bio-based plastics and natural fiber composites during co posting, anaerobic digestion and long-term soil incubation. *Polym. Degrad. Stab.* **2016**, *98*, 2583–2593. [CrossRef]
34. Rodriguez, E.J.; Marcos, B.; Huneault, M.A. Hydrolysis of polylactide in aqueous media. *J. Appl. Polym. Sci.* **2016**, *133*. [CrossRe
35. Simmons, H.; Kontopoulou, M. Hydrolytic degradation of branched PLA produced by reactive extrusion. *Polym. Degrad. St* **2018**, *158*, 228–237. [CrossRef]
36. Momeni, S.; Ghomi, E.R.; Shakiba, M.; Shafiei-Navid, S.; Abdouss, M.; Bigham, A.; Khosravi, F.; Ahmadi, Z.; Faraji, M.; Abdou H.; et al. The Effect of Poly (Ethylene glycol) Emulation on the Degradation of PLA/Starch Composites. *Polymers* **2021**, *13*, 101 [CrossRef]
37. Tsuji, H.; Ikada, Y. Properties and morphology of poly(l-lactide) 4. Effects of structural parameters on long-term hydrolysis poly(l-lactide) in phosphate-buffered solution. *Polym. Degrad. Stab.* **2000**, *67*, 179–189. [CrossRef]
38. Tsuji, H.; Ikarashi, K. In vitro hydrolysis of poly(l-lactide) crystalline residues as extended-chain crystallites: III. Effects of pH ar enzyme. *Polym. Degrad. Stab.* **2004**, *85*, 647–656. [CrossRef]

coatings

icle

einforcement of Oracle Bones Using a Novel Silicone Coupling eagent for Preservation

ng Gao [1,*,†], Qiyu Liu [1,†], Peipei Wang [2] and Yongge Liu [1]

1 Key Laboratory of Oracle Bone Inscriptions Information Processing, Ministry of Education, Anyang 455000, China
2 School of Chemistry and Chemical Engineering, Anyang Normal University, Anyang 455000, China
* Correspondence: gaof@aynu.edu.cn
† These authors contributed equally to this work.

Abstract: Oracle bones are artifacts of great significance and value in the study of Chinese history and culture. Because of soil and bacterial erosion, bones become fragile, and the inscriptions on the surface become blurred, resulting in the loss of historical information on the bones. In early times, scholars often used adhesives for bonding and reinforcement, whereas in modern times, organic and inorganic materials have been used as reinforcement for oracle bones. In this study, the surface of oracle bone was protected and reinforced by a new silicone coupling reagent that could self-polymerize in the format of colorless solution with good antimicrobial properties. The DESPMA was applied to the surface of oracle bone by dropwise addition and curing, effectively protecting it from bacteria and slowing down the yellowing process. The results showed that the reagent could significantly improve the antimicrobial properties of bone samples and reduce the yellowing and discoloration caused by bacterial attack. The reagent barely affected the appearance of the bone samples. These findings are promising and valuable for effective application in bone protection and utilization.

Keywords: conservation; oracle bones; heritage preservation; antimicrobial properties; restoration of relics

1. Introduction

A large number of bone products exist in historical artifacts, most of which are made from ivory, tortoiseshell, and animal bones. Artifacts made from tortoiseshell and animal bones are known as oracle bones. The bones used in oracle bones include pig bones, sheep bones, bovine bones, and a few human bones.

The writing recorded on oracle bones provides important evidence for studying history, but the process of making oracle bones is complicated. Ancient peoples cauterized and drilled these bones and applied writing to them, which documented the lives of the nobles of the Shang Dynasty [1–3]. This action makes oracle bones fragile and difficult to preserve [4,5].

Research has shown that the chemical composition of oracle bones includes both inorganic and organic parts [6]. Organic materials account for 30%, mainly including proteins and polysaccharides, and inorganic materials account for approximately 70%, mainly comprising calcium phosphate, carbonate, and fluoride. After thousands of years of soil burial, the non-apatite portion of the mineral composition becomes more unstable, and some inorganic materials are dissolved [7,8]. This results in varying degrees of cracking, fragmentation, and yellowing of oracle bones.

After many years of soil burial, some chemicals, bacteria, and fungi in the soil again change the morphology and chemical composition of oracle bones [9–11]. Figure 1 shows the oracle bone ZR044547 from a collection in Taiwan, China, which was embellished and pasted in 2010 [12]. The images show that, as a result of the corrosion of the soil (Figure 1A), the color distribution of the surface is uneven, and some of the oracle bones have begun to show a darker brown color.

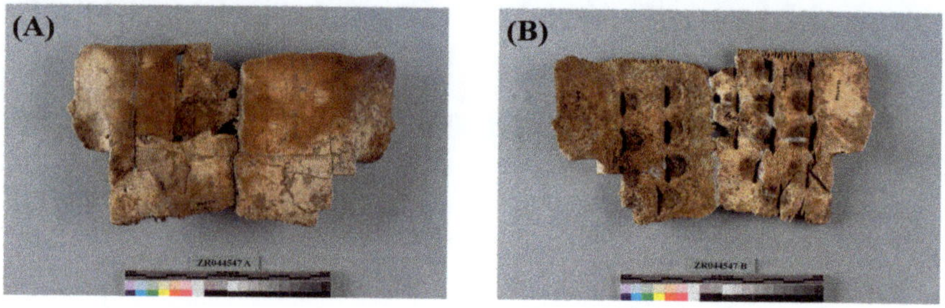

Figure 1. Cloth-covered glue marks on oracle bone ZR044547 on the front (**A**) and the back (**B**).

However, this information holds great significance to the study of history, which provides rich information on rituals, diet, and diseases [13–16]. In order to protect the bones, scholars began to study the use of chemicals to protect and repair oracle bones [17].

The physical preservation of the oracle bone is mainly achieved through vacuum storage to protect the oracle bone [18]; Because some oracle bones are fragile and cannot be vacuum treated, scholars began to study the use of chemical adhesives. In previous studies, plaster or glue adhesives were usually used to repair the bones. Glue adhesives were made mainly from cellulose nitrate and cellulose acetate dissolved in acetone [19]. Figure 1B shows a large trace of glue.

In general, the protective agent needs to have good working properties and provide cohesion, strength, and durability [17,20]. In practice, scholars have tried to use a variety of materials for the protection of oracle bones, such as B-72 acrylic resin, polyvinyl acetate, polyvinylidenebutyl alcohol, polyethylene glycol, and acrylic reinforcement [21–23]. They also have adopted different techniques for the use of these reagents, such as dripping, spray painting, submerging, and decompression infiltration [24,25].

In the 1980s, a class of organic silicon anti-weathering materials, including the compounds of methyl trimethoxysilane, polysiloxane, and silicone were developed [26–29]. Polymer resin materials have been widely used in the restoration of cultural relics, such as type 509 epoxy resin which is often used in porcelain, and three, a resin used in metal cultural relics coating. At the same time, 2-hydroxyethyl methacrylate (HEMA), as the main monomer component in composite resin material matrix, is also widely used in the field of dental materials (self-etch adhesives and luting composites) [30,31]. Researchers are also exploring the antioxidant and antibacterial properties of composites made with HEMA as a monomer.

In this study, we proposed a novel organic silicon coupling reagent protectant for the prevention of bacterial attack on fragile oracle bones. Figure 2 schematizes the mechanism of action of the proposed reagent on the oracle bone. The first image shows a large gap inside the oracle bone, the second image shows the addition of reagents to fill the gap, and the third image shows the inhibitory effect on bacteria. Our goal was to reduce the negative consequences of discoloration and corrosion of the oracle bone after bacterial attack and to reduce preservation damage. We evaluated the antimicrobial effect of the agent with X-ray diffraction (XRD), scanning electron microscopy (SEM), Vickers hardness measurement (n = 2, 3 readings per specimen), antimicrobial properties, and penetration velocity measurements.

Figure 2. Demonstration of reagent mechanism.

2. Materials and Methods

2.1. Reagents and Materials

2-Hydroxyethyl methacrylate (HEMA), saturated sodium chloride, and anhydrous sodium sulfate were purchased from National Pharmaceutical Group Co. (Beijing, China). Gelatin with the texture of colorless or light yellow flakes or lumps was produced by Zhonglian Chemical Reagent Co., Ltd. (Tianjin, China). DESPMA (2-(2-(Methacryloyloxy) ethyl)disulfide)ethyl(3-(diethoxymethylsilyl)propyl)carbamate) was synthesized according to a published study [32].

HEMA-CDI (10.07 g, 44.9 mmol) and 3-(diethoxymethylsilyl) propylamine (9.04 g, 47.24 mmol) were placed into a 100 mL round bottom flask, then 50 mL of dry dichloromethane (DCM) was added. After stirring the reaction mixture for 1 h, the DCM was removed and the residual oily substances were precipitated in petroleum ether twice. The obtained light yellow oily substance was dried under vacuum (yield: 92.9%) [33].

Real oracle bones are precious and their chemical composition would be unpredictably damaged by their use in this study. In general research, real oracle bones are not usually allowed to be used for experiments. According to the scientific experimental method, we used two materials in this experiment: cauterized bovine bone and simulated bone samples. The cauterized bovine bone simulated the production process of the oracle bone, and the simulated bone sample simulated the composition of the oracle bone [34].

2.2. Bovine Bone Samples

It has been found that the oracle bones burn at different temperatures and, if soaked with acid after excavation, would be weathered [35]. High temperatures, drilling, and chiseling operations will change the physical structure and chemical composition of oracle bone [36]. To better verify the effect of the reagent, we used cauterized bovine bone in the experiment. To prepare the bovine bone, the following steps were used:

1. The bones were boiled for 1 h in a solution of sodium bicarbonate to remove oil and grease;
2. The cooled bones were dried in the air for 2 d and then sawed into rectangles of similar size by using a Low Speed Diamond Saw (SYJ-150, Shanghai Optical Instrument Factory, Shanghai, China);
3. The cut bones were placed in a muffin pan, burned at 550 °C for 4 h and then cooled for 8 h;
4. The resulting bones were soaked in 4.5% hydrochloric acid for 5 min, rinsed with water three to four times, and then dried in the air for 2 d.

Figure 3 shows the bovine bone samples before (Figure 3A) and after being treated by boiling in sodium bicarbonate solution and drying at room temperature (Figure 3B), and by firing at high temperature (550 °C) (Figure 3C). The treated bone exhibited a whitening effect, a decrease in bone quality and surface curvature, and/or a tendency to release white powder on contact. These negative effects seriously affected the integrity of the bone surface.

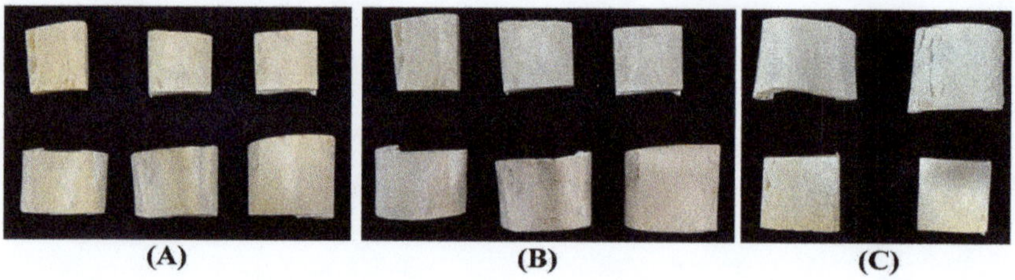

Figure 3. (**A**) Bovine bone samples. (**B**) Bovine bones boiled and dried. (**C**) Bovine bones fired at 550 °C.

Figure 4 shows the XRD results of the real oracle bone and bovine bone samples. When the real oracle bone was tested, the diffraction peaks of gelatin appeared at 25.91°, 31.84°, 39.49° and 48.59° (Figure 4A), which was consistent with the hydroxyapatite diffraction data in the database. The bovine bone sample (Figure 4B) showed diffraction peaks that almost identical to those of the real oracle bone.

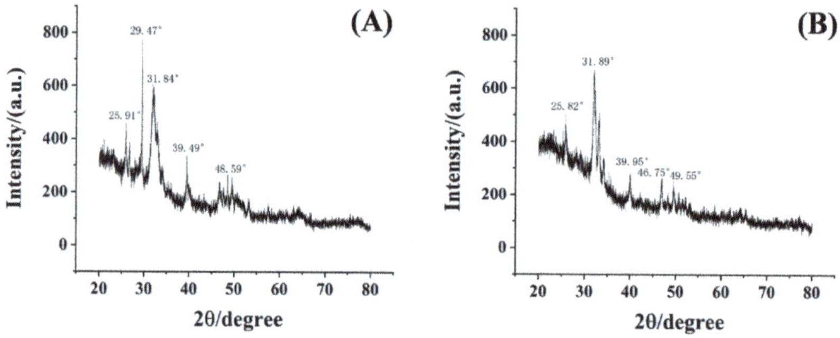

Figure 4. XRD results of real oracle bone (**A**) and bovine bone (**B**) samples.

2.3. Simulated Bone Samples

The composition of oracle bone will change after thousands of years of burial. To better verify the effect of the reagent, we used a simulated bone sample in the experiment. The simulated bone samples were prepared based on the percentage content of different metacarpal bone substances, as shown in Table 1 [5]. Then, 0.23 g of water, 0.79 g of gelatin, 0.3 g of calcium carbonate, 1.3 g of undried and unsterilized earth, and 6.97 g of hydroxyapatite were mixed. The mixture was stirred in an agate mortar and divided into five equal portions of ~2 g per sample. Each portion was pressed into a powder tablet at 1 MPa by using SZT-15T manual powder tablet press (Tianjin Zhongtuo Science and Technology Development Co, Tianjin, China).

Table 1. Percentage chemical composition of oracle bone substances.

Water	Organic Matter	Carbonate	Soils	Hydroxyapatite
2.40	8.24	3.13	13.56	72.68

Silica resin contains a hydrophobic alkyl group and a silicon oxide group for the formation of the Si-O-Si network. It has the flexibility of organic materials, water resistance, and the aging resistance of inorganic materials. Research shows that the silicone resin has good permeability and combined strength to strengthen the internal structure of the overall mechanical strength and aging resistance of the relics.

Scanning electron microscopy (SEM, FEI SIRION-100, 5.0 kV of accelerating voltage, Hillsboro, OR, USA), Fourier transform infrared spectroscopy (FTIR, Thermo Scientific™ Nicolet™ iN10, 4000–450 cm^{-1} range, Thermo Fisher Scientific, Madison, WI, USA), and X-ray diffractometry (XRD, AXS D8 ADVANCE, Cu Ka radiation, scan range 2θ = 10–80°, Bruker, Billerica, MA, USA) were used to analyze the structure and composition of the samples.

2.4. Plate-Coated Colony Counting Assay

The bovine bone samples treated by DESPMA were used as carriers for the autoclaved materials. The antimicrobial properties of the samples were verified using colony counting on agar plates. The experimental samples and equipment were sterilized before the experiment. All the samples were sterilized by ultraviolet disinfection, and the sterilized samples were put into disposable plates to characterize the antibacterial properties of the surface.

1. 500 g of soil was collected, soaked in deionized water, covered with the soil surface for 2–5 cm, stirred and settled overnight, and an appropriate amount of supernatant was taken as the source of the flora for this characterization experiment.
2. Sample A, treated by DESPMA, had a solution of soil dropped onto on the surface. As the liquid needs to wrap around the sample surface as much as possible, the liquid was spread as far as possible.
3. The sample B, without the DESPMA, had a solution of soil dropped onto the surface. As the liquid needs to wrap around the sample surface as much as possible, the liquid was spread as far as possible.
4. The incubation temperature was 37 °C, the humidity was 60%–98%, and the incubation time was 24/48/72 h.
5. After incubation, 10 (L medium) was taken from the surface of samples A and B. The solutions were coated in three gradients and counted separately, namely, 1×, 10× and 100×.

3. Results

To improve the precision of the permeability at the molecular level, a new component, a reducible silica monomer called DESPMA, was used. The molecular structure of the reagent along with its ^{1}H nuclear magnetic resonance spectroscopy (^{1}H NMR) data are shown in Figure 5. We perform integration on specific peaks during the process. The integrated areas of the two are in a ratio of 4:6, which demonstrates that the two raw materials are coupled in a 1:1 ratio. The diethoxymethylsilyl property of DESPMA by in situ crosslinking caused a lower crosslink density and a greater permeability [32].

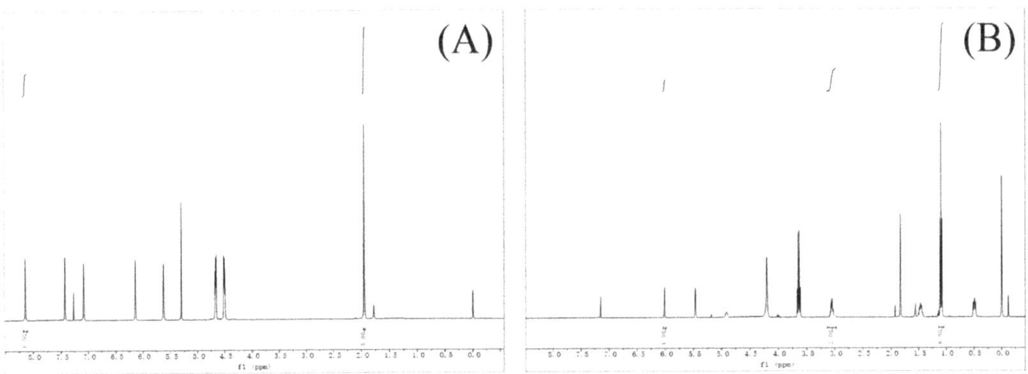

Figure 5. (**A**) ^{1}H NMR spectrum/spectra of HEMA-CDI. (**B**) ^{1}H NMR spectrum/spectra of DESPMA.

The methacrylate group in this monomer was used for free radical polymerization, the disulfide bond has a certain reduction response, and the two alkoxygen groups connected to the silicon atoms were used for crosslinking curing.

As shown in Scheme 1, DESPMA first penetrated into the oracle bone, and then triggered the crosslinking reaction, playing a role in reinforcement and protection.

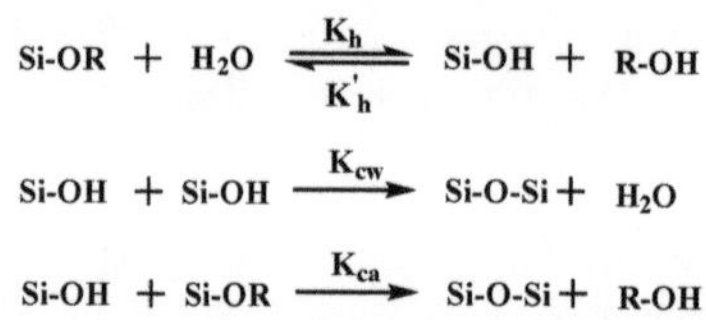

Scheme 1. Crosslinking reactions of DESPMA.

From the infrared spectrum of DEPMA, it can be observed that, the stretching vibration absorption peak of N-H is at 3340 cm^{-1}, the stretching vibration absorption peaks of C-H are at 2973 cm^{-1} and 2881 cm^{-1}, the stretching vibration absorption peak of the ester group C = O is at 1719 cm^{-1}, the stretching vibration absorption peak of the C = C double bond is at 1527 cm^{-1}, the bending vibration absorption peak of-CH_2-is at 1444 cm^{-1}, the bending vibration absorption peak of-CH_3 is at 1390 cm^{-1}, the stretching vibration absorption peak of C-O is at 1244 cm^{-1}, the in-plane bending vibration absorption peak of C-H is at 1164 cm^{-1}, the out-plane bending vibration absorption peak of C-H is at 942 cm^{-1}, and the antisymmetric stretching vibration absorption peaks of Si-O are at 1072 cm^{-1} and 763 cm^{-1} (Figure 6).

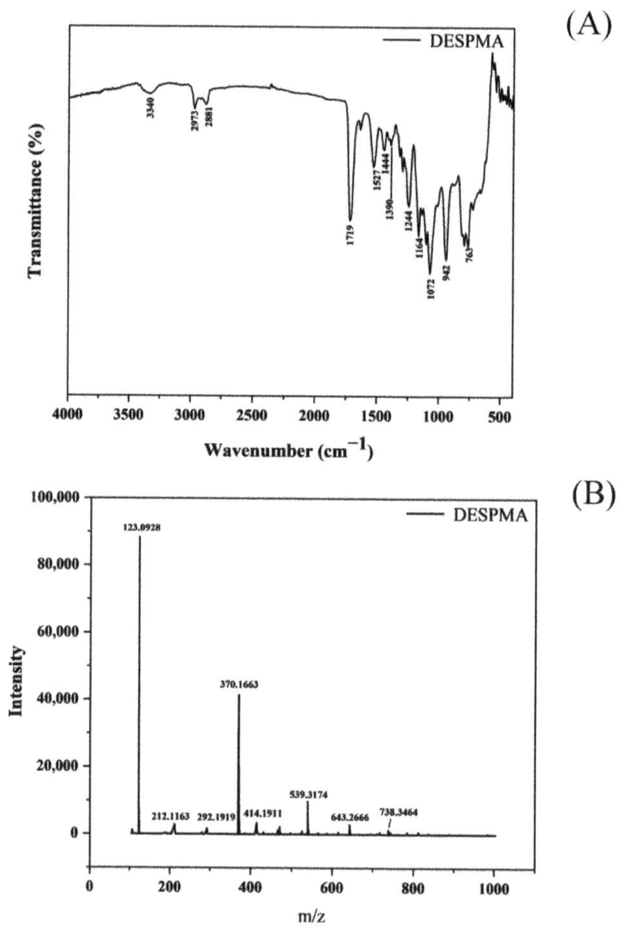

Figure 6. (**A**) The FTIR results of the DESPMA. (**B**) The mass spec results of the DESPMA.

The penetration ability of the reagent is a key factor for the effect of the reagent, and the penetration ability determines the area of the reinforced oracle bone.

In order to better improve the permeability, the density of Si-O-Si bonds was reduced, and the silicon atoms act in only two directions that can be connected to the crosslinked network. The remaining two directions are not connected to the crosslinked network. The introduction of silicon oxide groups (Si-O-Si) can change the size of the pores in the network. By adjusting the number and distribution of silicon oxide groups, the pore size precisely controls the permeability of the material. Larger pores favor the diffusion and penetration of macromolecules or hydrophobic molecules.

3.1. Micro-Structure Evolution of the Simulated Bone Samples

Scanning electron microscopy (SEM) was utilized to investigate the evolution of the microstructure of the samples. The magnification was adjusted to 200 μm and 5.00 μm to meticulously analyze the surface features. As shown in Figure 7A,C, the untreated samples present a rough and porous surface texture. In contrast, the samples treated by DESPMA (Figure 7B,D) exhibit a smoother surface. The reagents fill in the pore to some extent, thereby reinforcing and protecting the sample.

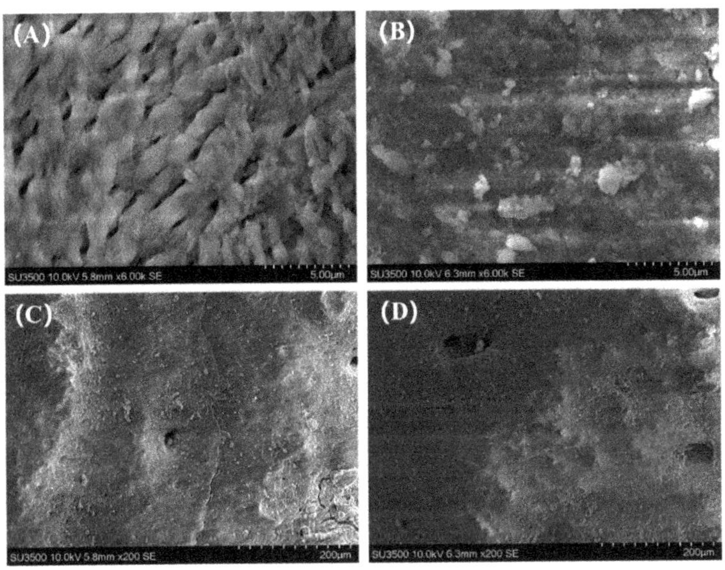

Figure 7. The micro-structure evolution of the simulated bone samples. (**A**) Untreated samples were amplified to 5.00 μm. (**B**) Samples treated by DESPMA were amplified to 5.00 μm. (**C**) Untreated samples were amplified to 200 μm. (**D**) Samples treated by DESPMA were amplified to 200 μm.

3.2. Surface Hardness Change

The reinforcement performance of the reagent was measured using Vickers hardness. For two samples measuring 2 cm × 2 cm, one was untreated, and the other one treated by DESPMA. The specimens were placed in a microhardness indenter (Micromet 5100, Buehler, Singapore). Three indents were made in each specimen, near the center of the specimen and at least 0.5 mm away from each other. As shown in Figure 8, the hardness of sample b increased by 6.8% to 26.8%. To increase the reliability of the experiment, by averaging the three values, the hardness of sample B was increased by 14.3%.

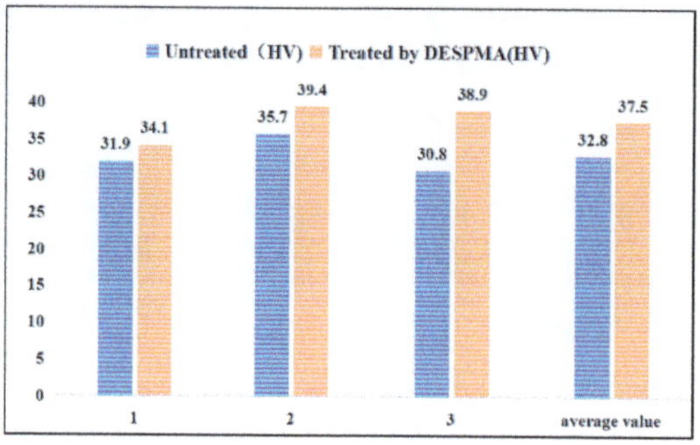

Figure 8. The surface hardness change in the simulated bone samples.

3.3. Consolidation of Reagents

The consolidation of the adhesive is a key factor in oracle bone reinforcement. Because of their low boiling point, all solvents used in this article can evaporate in an open environment at room temperature. In addition, the consolidation speed can affect the applicable range and condition of the material to a large extent.

Two reagents, B72 (acrylic resin adhesive) and PS (liquid sodium silicate) were experimentally tested and compared with the present reagent. As shown in Figure 9, to make the experiment comparable, all materials were applied to the bone sample surface only once, and the mass changes were recorded immediately after the application. Reagent addition was followed by rapid volatilization and then a slow consolidation reaction. The consolidation rate for DESPMA (Figure 9C) is slightly lower than that of B72 (Figure 9B) but higher than that of PS (Figure 9A).

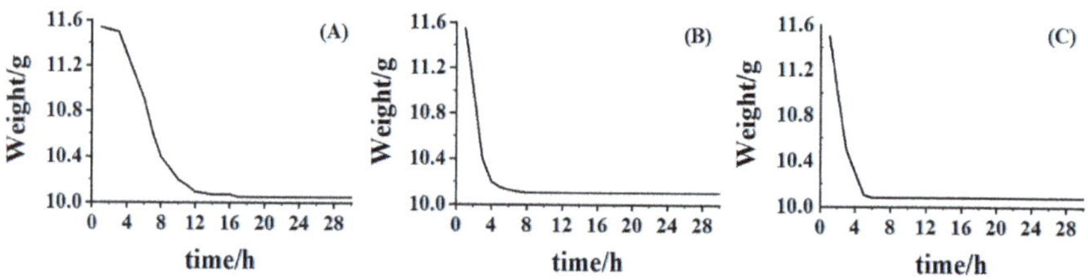

Figure 9. Consolidation rates of different solvents. **(A)** 1.5% PS aqueous solution. **(B)** 1.5% B72 acetone solution. **(C)** 1.5% DESPMA dichloromethane solution.

To investigate the possible mechanism of the change above, the composition of the bone samples was monitored during the consolidating process by FTIR analysis. The infrared spectrogram of the bovine bone sample reveals distinct absorption peaks corresponding to specific vibrational modes. At 1413 cm^{-1}, an asymmetric stretching vibration absorption peak is attributed to the carbonate ion (CO_3^{2-}). Similarly, at 1028 cm^{-1}, an asymmetric stretching vibration absorption peak is indicative of the P-O bond. The symmetric stretching vibration of the P-O bond is observed at 962 cm^{-1}, while the symmetric stretching vibration of the carbonate ion (CO_3^{2-}) is discernible at 873 cm^{-1}. Furthermore, the stretching vibration of the Ca-O bond is characterized by an absorption peak at 631 cm^{-1}.

The N-H group (at 3340 cm^{-1}) within DESPMA may engage in hydrogen bonding with the phosphate (P-O) or carbonate (CO_3^{2-}) groups present in the bovine bone. This interaction can enhance the cohesion and increase the hardness of the sample, without generating any new chemical absorption peaks in the infrared (IR) spectrum. The Si-O groups in DESPMA (1072 cm^{-1} and 763 cm^{-1}) may form a silica bridging bond with the hydroxyl group or water molecules in the bovine bone, which helps to improve the crosslinking density and stability of the material (Figure 10).

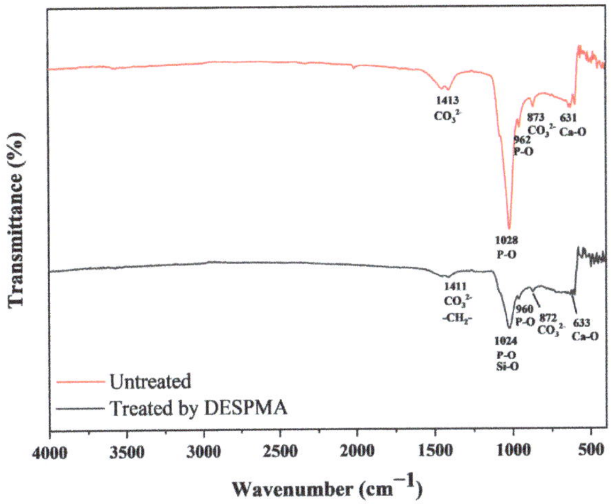

Figure 10. The FTIR results of the bone samples before and after the consolidation treatment.

3.4. Volatility of Reagents

The volatilization speed of the solvent in the reinforcement material has a great impact on the final protection effect. A fast volatilization speed may reduce the depth of the reinforcement agent, and even cause a reverse migration phenomenon of the reinforcement agent, thus resulting in secondary damage. Therefore, we tested the volatility of the solution.

We also used a circular simulated bone sample of the same diameter in this experiment. After complete coverage of the entire surface with the reagent, the size of the area covered by the liquid on the surface of the bone sample was determined and the time it took for the reagent to completely vaporize was recorded. The bone sample had a diameter of R and an area of πR^2, and the reagent covered an area with a diameter of r and an area of πr^2. The percentage of the area covered, E, was calculated as follows in Equation (1):

$$E = r^2/R^2, \tag{1}$$

As shown in Figure 11, DESPMA evaporated faster when the reagents were applied to the bone surface, avoiding chemical damage to the surface of the oracle bone to a greater extent and minimizing influence on surface color and appearance. The curve will eventually become zero, since the solvent of the reagent is volatile.

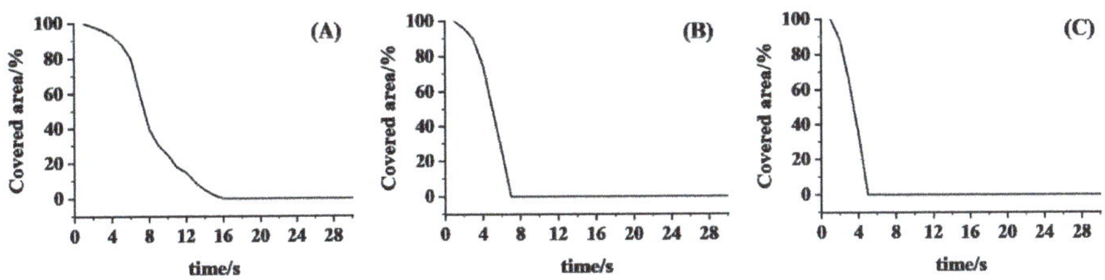

Figure 11. Comparison of the degree of volatilization of solvents. (**A**) 1.5% PS aqueous solution. (**B**) 1.5% B72 acetone solution. (**C**) 1.5% DESPMA dichloromethane solution.

3.5. Reagent Protection

Heat resistance is one of the important performance indexes of silicone coating. We also performed another test by dropping the same solution on a slide coated with hydroxyapatite gelatin and placing it in an oven at 40 °C. The breakage and peeling of the slides were recorded. In this experiment, the hydroxyapatite gelatin solution was added dropwise to the slides and it solidified to form a white solid-mounted flake covering the entire slide; then, 2 mL of water, PS, B72, and DESPMA solutions were added dropwise to each of the four slides. After dying at room temperature, the slides were heated in an oven at 40 °C. The time until the cracking of hydroxyapatite gelatin was recorded. As shown in Figure 12, the sample treated with an aqueous solution cracked after only 15 min of heating in the drying oven. The protection time was extended to 34 min for the sample treated with B72 solution, 45 min for the sample treated with PS solution, and 48 min for the sample treated with DESPMA solution. This result can be attributed to the high bond energy of the Si-O bond (~460 kJ/mol) [37] that endows the reagent with excellent heat resistance and chemical stability.

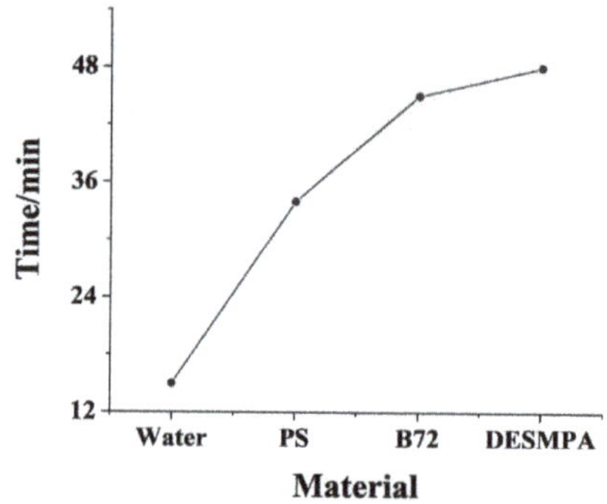

Figure 12. Time of appearance of cracking for the samples on slides.

3.6. *Effects of Bacteria on Agar Plates*

Figure 13 shows the petri dish coated with the solution of soil. Figure 13B shows a yellow bacterial colony, identified as *Staphylococcus aureus*. The next day, we observed multiple colonies with white and yellow dots on the petri dishes, indicating a relatively high concentration of microorganisms in the solution.

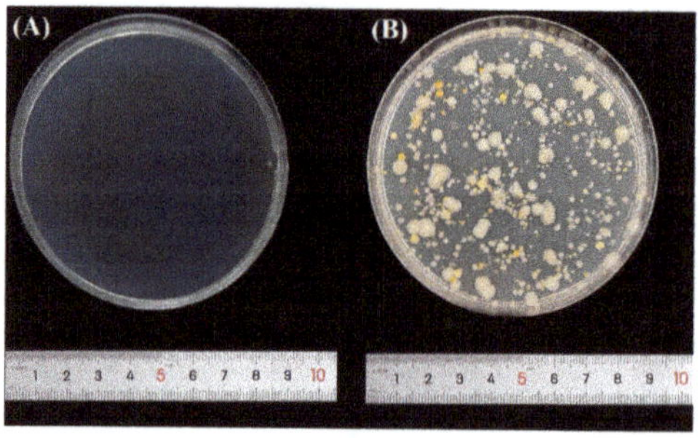

Figure 13. Cell culture medium. (**A**) The original petri dish. (**B**) The treated petri dish.

Samples were taken every 24 h and coated at a multiple dilution of 1, 10 and 100, where (a) (b) (c) is the sample treated by DESPMA, and (d) (e) (f) is the untreated sample. Figure 14A shows that the number of colonies on ×100-fold diluted (c) plates was 461 and on (f) plates was 1121, decreasing by 58.8%. Figure 14 shows that DESPMA inhibited bacterial growth to some extent.

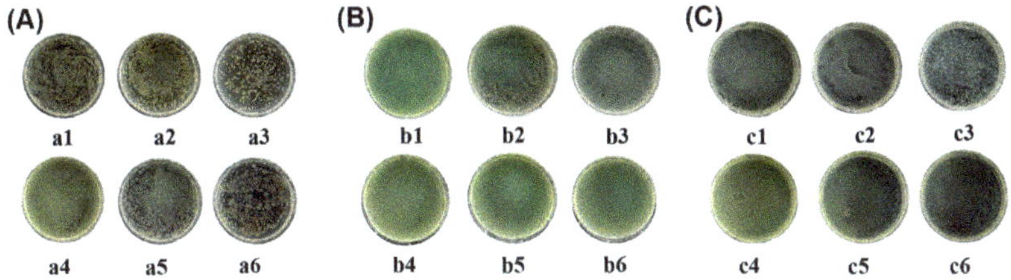

Figure 14. (**A**) The 24-h culture medium was sampled and coated. (**B**) The 48-h culture medium was sampled and coated. (**C**) The 72-h culture medium was sampled and coated.

3.7. Effects of Bacteria on the Bone Samples

The simulated bone samples and bovine bone samples treated by DESPMA were used as carriers for the autoclaved materials. A series of tests was conducted to determine the effectiveness of the different solutions in preventing yellowing and damage of the bovine bone and simulated bone samples. To do this, we pipetted 2 mL each of B72, PS, and DESPMA solutions onto the samples and allowed them to cure. The chromatographic solutions were then applied uniformly to the surface and any changes in appearance were recorded, The numbers of yellow patches and microflora on the samples were examined by microscopy and counted. This rigorous procedure ensured the accuracy and reliability of our results.

As the results in Tables 2 and 3 show, the bone samples treated by DESPMA developed fewer bacterial colonies. Using day 28 as a reference, there was the lowest number of colony counted on the surface of the bone samples treated by DESPMA, and a relatively good performance in both the simulated bone samples and the bovine bone samples. The degree of yellowing on the bovine bone samples increased only slightly compared with the simulated bone samples, highlighting the effectiveness of the DESPMA solution in preventing yellowing.

Table 2. Bacterial colonies counted on simulated bone samples.

Solution	Day 1	Day 7	Day 14	Day 21	Day 28
DESPMA	0	0	0	2	4
PS	0	0	2	5	6
B72	0	0	2	4	5
Water	0	0	3	5	8

Table 3. Bacterial colonies counted on for bovine bone samples.

Solution	Day 1	Day 7	Day 14	Day 21	Day 28
DESPMA	0	0	2	3	5
PS	0	3	5	7	8
B72	0	2	4	6	8
Water	0	3	3	7	9

4. Conclusions

The physical protection of oracle bones is crucial. In this study, we proposed the use of a chemical method to protect the surface of oracle bones from erosion. Specifically, we successfully developed a new silicone resin reagent (DESPMA). The resulting product was then dissolved in dichloromethane to form a reinforcer solution with superior antimicrobial and penetration properties. We applied droplets of the solution to the surface of the oracle bones to test its effectiveness as a protective material. After the curing treatment, the bone samples showed improved protection and antimicrobial properties, and the change in appearance was within acceptable limits. This study proposed a new means to protect oracle bones, although the reagents used should be further refined so that they can provide better protection for the information on the bone surfaces by strengthening the internal physical structure of oracle bones.

Author Contributions: Conceptualization, F.G. and Q.L.; methodology, F.G.; data curation, Q.L.; investigation, P.W.; writing—original draft preparation, Q.L.; writing—review and editing, F.G., P.W. and Y.L.; project administration, F.G. All authors have read and agreed to the published version of the manuscript.

Funding: This research was funded by the National Natural Science Foundation of China (No. 62106007), Sub-project of major program of the National Social Science Fund of China (No. 20&ZD305), Henan Revitalization Cultural Engineering Special Project (No. 2023XWH296), Henan Province Science and Technology Research Project (No. 232102320169), Henan Provincial Colleges and Universities Youth Key Teacher Training Plan (No. 2021GGJS129).

Institutional Review Board Statement: Not applicable.

Informed Consent Statement: Not applicable.

Data Availability Statement: Data is contained within the article.

Conflicts of Interest: The authors declare no conflicts of interest.

References

1. Wang, C.; Yang, Q.; Liu, L.; Yang, J.; Zhou, W.; Dang, X. Status of research on the conservation of bone and antler relics. *Sc Conserv. Archaeol.* **2016**, *1*, 118–125.
2. Du, A.; Ma, Y. Conservation treatment of Yinxu oracle bone inscriptions. *Cult. Relics Cent. China* **2007**, *6*, 87–90.
3. Flad, R.K. Divination and power: A multiregional view of the development of oracle bone divination in early China. *Cur Anthropol.* **2008**, *49*, 403–437. [CrossRef]
4. Qiao, R.; Yang, L.; Pang, K. Making visual sense of oracle bones for you and me. In Proceedings of the IEEE/CVF Conference o Computer Vision and Pattern Recognition (CVPR), Seattle, WA, USA, 17–21 June 2024; pp. 12656–12665.
5. Wang, C.; Bai, C.; Yang, J.; Wang, Y. Study on the conservation status of Zhouyuan oracle bones. *Sci. Conserv. Archaeol.* **2015**, *1*, 003
6. Lemaire, A.; Varin, J.; Lamret, F. Staphylococcus aureus behavior on artificial surfaces mimicking bone environment. *Pathogen* **2023**, *12*, 384. [CrossRef]
7. Wu, Y.; Lan, J.; Wu, M. Processing, microstructure, and performance of robocast clay-based ceramics incorporating hollo alumina microspheres. *Materials* **2024**, *17*, 1603. [CrossRef]
8. Chen, X.; Yuan, S. Conservation of Zhouyuan oracle bone inscriptions. *J. Natl. Mus. China* **2002**, *000*, 81–87.
9. Pérez, L.; Sanchis, A.; Hernández, C.M.; Galván, B.; Sala, R.; Mallol, C. Hearths and bones: An experimental study to explor temporality in archaeological contexts based on taphonomical changes in burnt bones. *J. Archaeol. Sci. Rep.* **2017**, *11*, 287–30 [CrossRef]
10. Ullmann, P.V.; Schweitzer, M.H. A statistical meta-analysis of lithologic and other potential controls on fossil bone cellular an soft tissue preservation. *Palaios* **2023**, *38*, 246–257. [CrossRef]
11. Kibblewhite, M.; Tóth, G.; Hermann, T. Predicting the preservation of cultural artefacts and buried materials in soil. *Stoten* **201** *529*, 249–263. [CrossRef]
12. Yuan, S.; Wu, X.; Liu, K.; Guo, Z.; Cheng, X.; Pan, Y.; Wang, J. Removal of contaminants from oracle bones during sampl pretreatment. *Radiocarbon* **2007**, *49*, 211–216. [CrossRef]
13. Liu, Y.; Lu, R.; He, L.; Wang, X.; Wang, L.; Lv, X.; Zhang, K.; Yang, F. Consolidation of fragile oracle bones using nano calciur sulfate hemihydrate as a protectant. *Coatings* **2022**, *12*, 860–871. [CrossRef]
14. Zhao, X.; Tang, J.; Gu, Z.; Shi, J.; Yang, Y.; Wang, C.J. Investigating the tool marks on oracle bones inscriptions from the Yinxu sit Henan province, China. *Microsc. Res. Tech.* **2016**, *79*, 827–832. [CrossRef]
15. Richter, A. Stories of coping with sickness: Illness narratives in early medieval Chinese anecdotal literature. *Chin. Med. Cult.* **202** *6*, 175–182. [CrossRef]
16. Liang, L.; Li, L. On diet and disease in oracle bone inscriptions. *Chin. Med. J.* **2023**, *44*, 77–80.
17. Hyeok, K. A study on the relationship between the period of oracle bone inscriptions and the development of character form *J. Chin. Writ. Syst.* **2023**, *7*, 86–89.
18. Xiao, L.; Sun, J. Study on temporary protection of ivories, buckteeth and antlers unearthed from Jinsha site. *Sci. Conserv. Archaeo* **2002**, *14*, 26–30.
19. Liu, Z. Preservation and scientific analysis of oracle objects: A preliminary study of oracle maintenance methods in the institut of history and language. *J. Cult. Herit. Conserv.* **2022**, *61*, 7–38.
20. North, A.; Balonis, M.; Kakoulli, I. Biomimetic hydroxyapatite as a new consolidating agent for archaeological bone. *Stud. Conser* **2016**, *61*, 146–161. [CrossRef]

Smith, D.R.; Martin, E.K.; Kaufman, B.L. The bottom line: Exploring analytical methods for assessing preservation in archaeological bone using ftir-atr. *J. Archaeol. Sci. Rep.* **2023**, *50*, 104014. [CrossRef]

Yang, L.; Zhang, J.; Huang, J. Cause analysis and application of surface deepening effect of paraloid b72 reinforcement agent for cultural relics. *West. Archaeol.* **2021**, *1*, 315–321.

Zheng, J. Application of acrylic resin in the field of cultural relics protection. *World Antiq.* **2018**, *06*, 74–76.

Li, Y.; Ling, X.; Yang, L.; Zhao, X.; Sun, M. Application of modern chemical materials in the protection of bone and keratin relics. *Polym. Mater. Sci. Eng.* **2021**, *37*, 8.

Du, A.; Zhou, S. Scientific analysis and restoration of several bronze artifacts. *Cult. Relics Prot. Archaeol. Sci.* **2004**, 46–51.

Wang, J. Application of organosilicon materials in the protection of stone cultural relics against weathering. *J. Longdong Univ. Nat. Sci. Ed.* **2020**, *31*, 5.

Wang, W.; Fan, M.; You, S. Study on the reinforcement of ancient brick wall cultural relics by silicone composite materials. *Guangdong Chem. Ind.* **2013**, 44–45.

Zhao, P.; Zhang, Y.S.; Shen, Y. Advancements in artificial hydraulic lime composites for sustainable restoration of stone cultural heritage. *Sci. Adv. Mater.* **2023**, *12*, 15. [CrossRef]

Xu, J.; Zhang, T.; Jiang, Y. Nano-silica/fluorinated polyacrylate composites as surface protective coatings for simulated stone cultural relic protection. *J. Appl. Polym. Sci.* **2022**, *139*, 52953. [CrossRef]

Moszner, N.; Hirt, T. New polymer-chemical developments in clinical dental polymer materials: Enamel–dentin adhesives and restorative composites. *J. Polym. Sci. Part A Polym. Chem.* **2012**, *50*, 4369–4402. [CrossRef]

Pimentel, R.; Paula, L.; Ribeiro, E. Evaluation of the bond strength of self-etching adhesive systems containing HEMA and 10-MDP monomers: Bond strength of adhesives containing HEMA and 10-MDP. *Int. J. Dent.* **2022**, *1*, 5756649.

Zhang, X.; Liu, F.; Li, X. The fabrication of hybrid micelles with enhanced permeability for drug delivery via a diethoxymethylsilyl-based crosslinking strategy. *J. Polym. Chem.* **2019**, *10*, 4529–4536. [CrossRef]

Zhang, X.; Wang, P.; Wang, X. Stabilized, ROS-sensitive β-cyclodextrin-grafted hyaluronic supramolecular nanocontainers for CD44-targeted anticancer drug delivery. *Colloids Surf. B Biointerfaces* **2024**, *242*, 114081. [CrossRef] [PubMed]

Barrett, T.H. Japanese monks and chinese books: Glimpses of buddhist sinology in early tokugawa japan. *Religions* **2021**, *12*, 871. [CrossRef]

Chen, X.; Yuan, S. Study on the burning condition and weathering causes of zhouyuan oracle bones. *Sci. Conserv. Archaeol.* **2004**, *16*, 20–28.

Liu, Y.; Zhang, Y.; Wang, Q.; Li, T.; Ma, W.; Yang, F.; Chen, L.; Zhao, D.; Yan, X. Consolidation of fragile weathered bone relics using hydroxyapatite material as consolidant. *J. Inorg. Mater.* **2023**, *38*, 1345–1354. [CrossRef]

Yilgör, E.; Yilgör, I. Silicone containing copolymers: Synthesis, properties and applications. *Prog. Polym. Sci.* **2014**, *39*, 1165–1195. [CrossRef]

MDPI AG
Grosspeteranlage 5
4052 Basel
Switzerland
Tel.: +41 61 683 77 34

Coatings Editorial Office
E-mail: coatings@mdpi.com
www.mdpi.com/journal/coatings

www.ingramcontent.com/pod-product-compliance
Lightning Source LLC
Chambersburg PA
CBHW062102220926
51850CB00010B/2861
* 9 7 8 3 7 2 5 8 6 0 1 1 1 *